ANALYTICAL ASPECTS OF TECHNETIUM SPECIATION IN SPINACH PLANTS

AF740317

The research described in this thesis was performed within the Department of Radiochemistry of the Interfaculty Reactor Institute, Delft University of Technology, Mekelweg 15, 2629 JB Delft, The Netherlands.

Het in dit proefschrift beschreven onderzoek is uitgevoerd in de afdeling Radiochemie van het Interfacultair Reactor Instituut van de Technische Universiteit Delft, Mekelweg 15, 2629 JB Delft.

ANALYTICAL ASPECTS OF TECHNETIUM SPECIATION IN SPINACH PLANTS

PROEFSCHRIFT

ter verkrijging van de graad van doctor
aan de Technische Universiteit Delft,
op gezag van de Rector Magnificus Prof. ir. K. F. Wakker,
in het openbaar te verdedigen ten overstaan van een commissie,
door het College voor Promoties aangewezen,
op dinsdag 17 maart 1998 te 13.30 uur
door

Arend Victor HARMS

doctorandus in de scheikunde

geboren te 's-Gravenhage

Dit proefschrift is goedgekeurd door de promotoren:
Prof. dr. ir. J. J. M. de Goeij
Prof. dr. ir. F. M. Everaerts

Samenstelling promotiecommissie:

Rector Magnificus, voorzitter
Prof. dr. ir. J. J. M. de Goeij, Technische Universiteit Delft, promotor
Prof. dr. ir. F. M. Everaerts, Technische Universiteit Eindhoven, promotor
Prof. dr. J. G. Kuenen, Technische Universiteit Delft
Prof. dr. H. Poppe, Universiteit van Amsterdam
Prof. dr. A. Verbruggen, Katholieke Universiteit Leuven, België
Dr. J. T. van Elteren, Technische Universiteit Delft
Dr. J. F. M. M. Lembrechts, Rijksinstituut voor Volksgezondheid en Milieuhygiëne

Dr. J. T. van Elteren en dr. H. Th. Wolterbeek hebben als begeleiders in belangrijke mate aan het totstandkomen van het proefschrift bijgedragen.

Published and distributed by:

Delft University Press
Mekelweg 4
2628 CD Delft
The Netherlands
Telephone: + 31 15 2783254
Telefax: + 31 15 2781661
E-mail: dup@dup.tudelft.nl

ISBN 90-407-1659-5 / CIP
NUGI: 813

Copyright © 1998 by Arend Victor Harms

All rights reserved. No part of the material protected by this copyright notice may be reproduced or utilised in any form or by any means, electronic or mechanical, including photocopying, recording or by any information storage and retrieval system, without written permission from the publisher: Delft University Press.

Printed in The Netherlands

Kiezen: niet wanneer men op de tweesprong staat,
maar wanneer men, veel verder gelopen, staan blijft
op het doodlopend pad, en zich herinnert.

Dan: terug? maar terug kan niet, tot hier is men gekomen.
Voorwaarts, voorwaarts, maar voorwaarts kan niet,
hier eindigt de weg.

Dan kan men alleen nog maar hier zijn vlag planten.

(Frida Vogels, De harde kern 3, met toestemming)

Aan mijn ouders

Table of Contents

7. Characterisation of Technetium Species Induced in Spinach Plants: 2) Combination of Selective Extraction and Chromatographic Techniques

8. Conclusions

List of Symbols

A	absorbance	AU
A	mass number	1
$E_{\beta^-,max}$	β^- particle energy	keV
E_γ	γ-ray energy	keV
ε	extinction coefficient	L mol^{-1} cm^{-1}
g	acceleration	m s^{-2}
ΔH°	standard enthalpy	kJ mol^{-1}
k'	capacity factor	1
K_d	distribution coefficient	1
K_{SEC}	size-exclusion distribution coefficient	1
K_{LC}	solute-surface interaction distribution coefficient	1
m	mass	g
M_r	molecular weight	1
R	gas constant	kJ K^{-1} mol^{-1}
$t_{1/2}$	half-life	y
V_0	column interstitial volume	mL
V_c	total column volume	mL
V_e	elution volume	mL
V_i	internal pore volume	mL
V_m	dead volume	mL
Z	atomic number	1

List of Abbreviations

cyclam	1,4,8,11-tetraazacyclotetradecane
Cys	cysteine
d	deuteron
DMSA	dimercaptosuccinic acid
DTPA	diethylenetriaminepentaacetic acid
EC	electron capture
EC	*N,N'*-1,2-ethylene-bis-cysteine
ECD	*N,N'*-1,2-ethylene-bis-cysteine diethyl ester
EDTA	ethylenediaminetetraacetic acid
EPR	electron paramagnetic resonance spectroscopy
EXAFS	extended X-ray absorption fine structure
GF-AAS	graphite furnace atomic absorption spectroscopy
Glu	glutamic acid
Gly	glycine
GSH	glutathione
HEMA	2-hydroxyethylmethacrylate ethylene dimethacrylate
HEPES	*N*-(2-hydroxyethyl)piperazine-*N'*-(2-ethanesulphonic acid)
HIDA	hepatic iminodiacetic acid
HM-PAO	hexamethylpropyleneamine oxime
HPLC	high-performance liquid chromatography
HR	high-resolution
ICP-MS	inductively coupled plasma mass spectrometry
IT	internal transformation
LPAS	laser induced photoacoustic spectrometry
LSC	liquid scintillation counting
MAG_3	mercaptoacetyl-triglycine
MDP	methylene diphosphonate

2-ME	2-mercaptoethanol
MeOH	methanol
MS	mass spectrometry
n	neutron
NADP	nicotinamide adenine dinucleotide phosphate
NMR	nuclear magnetic resonance spectroscopy
OAc	acetate
PCTFE	polychlorotrifluoroethylene
PEEK	polyether ether ketone
pen	penicillamine
RIMS	laser resonance ionisation mass spectrometry
RP-IPC	reversed-phase ion-pair chromatography
SEC	size-exclusion chromatography
TBA	tetrabutylammonium
Tc	technetium
TcO_4^-	pertechnetate
TcX	plant-induced technetium species
tetren	tetraethylenepentaamine
TFA	trifluoroacetic acid
Tris	tris(hydroxymethyl)aminomethane

Chapter 1

Scope of Thesis

1.1 INTRODUCTION

The radionuclide ^{99}Tc, an isotope of the element technetium, is produced in appreciable amounts by fission of ^{235}U and ^{239}Pu in the nuclear fuel cycle. The large fission yield combined with the long half-life and the considerable mobility in environmental systems of pertechnetate (TcO_4^-), the dominant chemical form of technetium in aerobic environments, has stimulated the interest in the radioecology of this element. Moreover, the remarkable ability of technetium, being an artificial element, to concentrate in marine and freshwater wildlife and terrestrial plants, which relates it to the human food chain, received much attention. The large transfer factor renders the possibility of using plants as bioindicator for environmental levels of technetium. To understand the behaviour of technetium in biological and environmental systems, it is necessary to obtain information on the chemical forms (species) of technetium involved. However, despite considerable progress made in the last 15 years, which started with the pioneering work of Lembrechts, Van Loon, and Desmet,* the bioavailability of technetium for plants and particularly the biochemical processes involved in the uptake, retention, transport and metabolism of technetium species in plants are still insufficiently unravelled. This thesis attempts to give analytical tools to make this type of research more

* References in Chapter 2

accessible by having a closer look at technetium speciation procedures in plants.

1.2 PROBLEM DEFINITION

The quantification of Tc species in plants is difficult for several reasons: i) species are unknown and ii) species might be altered as a result of the speciation procedure. These problems, however, are not specific for technetium, but do apply also to other elements. An additional difficulty is the extremely low "natural" level of technetium in the plant. Under extreme conditions "natural" Tc levels in plants may range up to approximately 0.2 μmol kg^{-1} dry weight, but generally these "natural" levels are much lower. Limitations of present procedures lie in the integrity of sample pretreatment methods, i.e., how can we isolate Tc species in an "interference-free" way from the plant, and in the availability of sufficiently assessed separation methods, i.e., how can we resolve as many species as possible with as little as possible interferences. In practice, this has resulted in operationally-defined speciation procedures without real checks of artefacts. A detailed look learns us that for the isolation of Tc species it is necessary to disrupt the plant cells by some kind of grinding, extrusion, cutting and slicing, sonication, immersion in liquid nitrogen followed by freeze-drying, or enzymatic disruption. Reduction, oxidation, hydrolysis, transchelation induced by the pretreatment procedures may cause profound changes to certain species. The multitude of organic and inorganic constituents in plants makes it difficult to predict exactly what problems might be encountered. Depending on the nature of the compounds to be isolated, various protective compounds might be added to the homogenisation medium. Inattentive addition of these protective compounds to the homogenisation medium, e.g., buffers, osmotica, metal ion chelators, detergents, reducing agents, may have an opposite effect. Furthermore, labile Tc species may dissociate and/or interact with components of the chromatographic system, i.e., stationary phase, mobile phase, and (metal) components of the equipment, resulting in adverse chromatographic behaviour. To reduce potential problems, "mild" separation conditions and the use of inert (titanium) or metal-free (PEEK) pump heads, injectors, tubing, fittings, and columns are required. In general, size-exclusion chromatography (SEC) is seen as a separation method with a low level of intrusion with regard to speciation alteration.

1.3 RESEARCH APPROACH

In this work, spinach, *Spinacea oleracea L.*, is used as a model plant for terrestrial systems, because of its relation to the human food chain, its ability to accumulate heavy metals, and certain experimental advantages, e.g., the relative ease of cultivation of the seedlings and the relative ease of maceration of its leaves. The role of technetium in this research with its chemical, environmental and analytical aspects is discussed in Chapter 2.

Chapters 3, 4, and 5, are dedicated to a closer look at the design of a reliable speciation method using SEC. Chapter 3 illustrates the role which a "protective" compound, i.e., 2-mercaptoethanol (2-ME), an inhibitor of polyphenol oxidase, might play in plant-Tc speciation studies. The reactivity of TcO_4^- and two Tc model species, i.e., anionic Tc-DTPA and cationic Tc-cyclam, with 2-ME is discussed. The Tc species are separated with low-pressure SEC, using Sephadex as stationary phase. It is shown that Tc-DTPA does react with 2-ME, presumably resulting in the formation of a $[Tc^{V}O\text{-}(2\text{-}ME)_2]$ complex.

In Chapter 4 the development of a "mild" high-performance SEC method for separation of Tc compounds is described. It is shown that non-size effects enhance the separation characteristics of this technique resulting in the separation of TcO_4^-, Tc-DTPA, and Tc-cyclam. Moreover, it is shown that the extent of these non-size effects has a close relation with the nature of the Tc species, temperature, ionic strength of the mobile phase, and nature of the stationary phase. One of the stationary phases, i.e., a copolymer of 2-hydroxyethyl methacrylate and ethylene dimethacrylate (HEMA), is also used in Chapter 5 to separate TcO_4^- from plant-induced Tc species (TcX).

Chapter 5 is dedicated to the development of an analytical procedure for Tc speciation in spinach plants using a dual Tc radiotracer technique. This technique provides corrections for procedural losses and redox conversions of both TcO_4^- and TcX. To this end, $^{99m}TcO_4^-$ or ^{95m}TcX are added during the analytical homogenization procedure of plants grown on nutrient solutions containing $^{99}TcO_4^-$, which was traced with $^{95m}TcO_4^-$ or $^{99m}TcO_4^-$, respectively. Care was taken to select a homogenisation medium without possible adverse additives. The buffer was selected on its useful pH range and its inertness, i.e., relative low metal ion binding capacity. No metal chelators, such as EDTA or sucrose, or reducing agents, such as 2-mercaptoethanol, were added to the homogenisation medium, to avoid the formation of new Tc species by transchelation or procedural reduction of pertechnetate, respectively.

Chapters 6 and 7 are dedicated to a further investigation of the speciation of TcX itself with extraction and chromatographic techniques.

In Chapter 6 the characteristics of two separation techniques, i.e., SEC and reversed-phase ion-pair chromatography (RP-IPC), with tetrabutyl-ammonium acetate as ion-pair reagent, are compared for separation of TcX into different sub-species, while in Chapter 7 with a combination of selective extraction, SEC, and RP-IPC (with trifluoroacetic acid as ion-pair reagent) more information on the identity of some TcX sub-species is obtained. By comparison of the chromatographic behaviour of TcX with that of in vitro synthesised Tc complexes, it is shown that Tc^V-cysteine is a major sub-species of TcX. Furthermore, in both chapters it is shown, that the increased separating possibilities of RP-IPC are obscured by adverse interactions of some Tc species with the silica backbone of the reversed-phase packing. On the other hand, some Tc species, e.g., Tc^V-cysteine and TcO_4^-, were inert to an acid mobile phase containing trifluoroacetic acid (pH 2.2), thus enabling separation with RP-IPC. The Tc-glutathione complex, which might be expected as a sub-species of TcX, was not detected. Furthermore, the analytical chemical results obtained in these chapters are discussed in relation to the metabolism and plant physiological effects of technetium.

Finally, a general discussion of the results obtained in this thesis is given in Chapter 8.

Chapter 2

Technetium

2.1 DISCOVERY OF ELEMENT 43

The existence of element 43, eka-manganese, was predicted on the basis of the Periodic Table published by Mendeleev in 1869.[1,2] Several attempts, which resulted in proposed names such as polinium, ilmenium, davyum, and lucium, failed to isolate this element from minerals and ores.[2,3] From 1904 to 1906, M. Ogawa attempted to isolate element 43, which he named "nipponium", from sulphides extracted from the radioactive material thorianite.[4,5] In 1925, Noddack and co-workers[6] claimed the discovery of the elements 43 and 75, which they named "masurium" (Ma) (Masuren, East Prussia) and "rhenium" (Re) (L. *Rhenum*, Rhine), respectively, in concentrates of the minerals sperrylite, gadolinite, and columbite. However, in spite of a number of reports in the period from 1925 to 1927 by the same group, their masurium claim was never generally accepted.[7] In 1937 Perrier and Segré[8,9] obtained element 43 [in fact, the isotopes ^{95m}Tc and ^{97m}Tc as ^{94}Mo (d,n) and ^{96}Mo (d,n) reaction products, respectively] as the first artificially produced element by irradiation of molybdenum with deuterons. The short-lived isotope ^{99m}Tc was discovered by Seaborg and Segré in 1939.[10] They concluded that the ground state ^{99}Tc, whose emitted radiation remained undetected, possessed a half life greater than 40 years. After discovery of uranium fission in 1939, element 43 was found as ^{99m}Tc among the fission products.[11] By the late forties nuclear reactors were producing macroscopic amounts of the isotope ^{99}Tc, which was finally detected in 1946-47 in fission products[12,13] and in neutron-irradiated molybdenum.[14] The name technetium (τεχνητοs), which means "artificial" in Greek, was accepted in 1947.[15]

2.2 NUCLEAR PROPERTIES OF TECHNETIUM ISOTOPES

Technetium ($Z = 43$; isotopes with A ranging from 87 to 111) is the lightest element without any stable isotopes.[16] The isotopes with longest half-lives are ^{97}Tc (2.6×10^6 y), ^{98}Tc (4.2×10^6 y), and ^{99}Tc, (2.13×10^5 y). Of these isoptopes, only ^{99}Tc can be obtained in appreciable amounts (see Section 2.5). The nuclear characteristics of the three isotopes of technetium used in this work, i.e., ^{95m}Tc, ^{99m}Tc, and ^{99}Tc are given in Table 2.1.

Table 2.1. Nuclear characteristics[16] of the isotopes of Tc used

	half-life	main mode of decay	most abundant γ-energy (keV)	most abundant β and electron energy (keV)
^{95m}Tc	61 d	EC, IT, β^+	204 (66%) 582 (33%) 835 (28%)	β^+ particles: 513 (0.2%) 717 (0.3%) electrons: 2-204 (125%)
^{99m}Tc	6.02 h	IT	141 (89%)	electrons: 2-140 (122%)
^{99}Tc	2.13×10^5 y	β^-	–	β^- particles: 292 (100%)

2.3 CHEMISTRY OF TECHNETIUM

Technetium (Tc) is the *4d*-block metal in group 7 of the Periodic Table. The ground-state electronic configuration of the Tc atom is $[Kr]4d^5 5s^2$. The chemical properties resemble more or less those of the homologous elements manganese and rhenium and, to a lesser degree, those of neighbouring elements molybdenum and ruthenium. These properties are most similar to the ones of rhenium, but distinctly different with respect to redox behaviour and substitution lability of the lower oxidation states.[17] The oxidation states of technetium range from –I to VII. Technetium has a wide variety of coordination behaviour in addition to the variety of oxidation states, and its chemistry is among the most complex of the transition elements.[18-32] Pertechnetate, $[Tc^{VII}O_4]^-$, is among the most important compounds formed by this element. It shows, unlike MoO_4^{2-}, little tendency to form isopolyanions.[33] TcO_4^- is a much weaker oxidising agent than MnO_4^-, but it is reduced by HCl.[33] The VI oxidation state of Tc is dominated by the $[Tc^{VI}N]^{3+}$, $[Tc^{VI}N(O)Tc^{VI}N]^{4+}$, and $[Tc^{VI}N(\mu\text{-}O)_2Tc^{VI}N]^{2+}$ cores, which contain the Tc≡N bond.[34] Apart from the $[Tc^{V}N]^{2+}$ core, the V oxidation state is dominated by the oxo $[Tc^{V}O]^{3+}$, the *trans*-dioxo $[Tc^{V}O_2]^{+}$, and the $[Tc^{V}O(O)Tc^{V}O]^{4+}$ cores.[35] The electronic structure of the $[Tc^{V}O]^{3+}$ core renders this kinetically inert in ligand substitution reactions.[36] A variety of monomeric Tc^{IV} and dimeric (bis) μ-oxo $Tc^{III}Tc^{IV}$ and $Tc^{IV}Tc^{IV}$ species have been reported.[18,19] Tc^{IV} (d^3) complexes are considered to be inert.[36] A considerable number of binuclear Tc^{II} and Tc^{III} compounds containing $[Tc_2]^{4+}$, $[Tc_2]^{5+}$, and $[Tc_2]^{6+}$ cores, as well as monomeric Tc^{II} and Tc^{III} species have been characterised.[18,19,37] A considerable variety of stable mostly six-coordinate Tc^{I} complexes such as Tc-sestamibi[38] in aqueous media have been prepared. An important Tc^{0} species with great synthetic utility is the $[Tc^{0}_2(CO)_{10}]$ dimer.[39,40] Tc^{-I}, which is only known to exist in the $[Tc^{-I}(CO)_5]^-$ form,[41] is the rarest oxidation state of Tc.

2.4 USE OF ^{99m}Tc IN NUCLEAR MEDICINE

Diagnostic nuclear medicine procedures are based on the tendency of the body to concentrate some chemical form of a particular positron or gamma ray emitting isotope in the organ of interest.[42] Diagnostic nuclear medicine relies to a large extent on the use of ^{99m}Tc because of its favourable nuclear properties (half-life 6.02 h, E_γ 141 keV), its availability with a high specific activity from a $^{99}Mo/^{99m}Tc$ generator,[43] the variety of chemical forms in which it can be converted (see Table 2.2), and its relatively low cost.[44] ^{99m}Tc has almost no beta emissions, and emits only low-energy Auger electrons. The development of a generator system and of a large variety of labelling kits for the preparation of radiopharmaceuticals has led to the wide spread use of ^{99m}Tc in nuclear medicine. The generator has been developed at the Brookhaven National Laboratory (USA) in 1957 and is now based on the adsorption of carrier-free ^{99}Mo in the form of MoO_4^{2-} on an Al_2O_3 column. TcO_4^- is eluted from the column with physiological saline (0.15 mol L^{-1} NaCl), while the parent $^{99}MoO_4^{2-}$ remains strongly retained. The kits generally contain a reducing agent (usually Sn^{II}) and a chelating agent. The biodistribution of the resulting ^{99m}Tc complex is strongly dependent upon its formal charge; for example, some cationic complexes accumulate in the heart but cannot pass the blood-brain barrier, whilst neutral complexes do not accumulate in the heart but can pass the blood-brain barrier.[45] The most important ^{99m}Tc radiopharmaceuticals (with donorset and references) used as imaging agent are given in Table 2.2.

2.5 ^{99}Tc AS A FISSION PRODUCT

"Primordial" technetium is considered extinct on earth.[59] However, negligible amounts are produced by naturally occuring neutron-induced fission of ^{235}U and spontaneous fission of ^{238}U in uranium bearing minerals[60] and by activation of molybdenum ores by cosmic-ray neutrons. The production and subsequent introduction into our environment has started approximately 55 years ago with releases from the nuclear fuel cycle and fallout from nuclear weapon testing programs.[61,62] The total production of ^{99}Tc up to the end of 1993 was estimated to be 49,000 TBq (78 ton).[63] ^{99}Tc is the only technetium isotope produced in appreciable amounts, because of its high fission yield for thermal neutron fission of ^{235}U (isobaric yield: 6.06%)[64] and its long half-life. Technetium isotopes having mass numbers greater than 100 have similar fission yields, but because of their short half-lives (less than 18 min) they are not found in substantial quantities. ^{99}Tc is also formed in high abundance from thermal neutron fission of ^{233}U (4.8%), ^{239}Pu (5.9%), and fast neutron fission of ^{239}Pu (5.9%), ^{238}U (6.3%), and ^{232}Th (2.7%).[65]

Table 2.2. ^{99m}Tc radiopharmaceuticals used in nuclear medicine

brain perfusion	$[Tc^{V}O\text{-d,l-HM-PAO}]$	ON_4	46
	$[Tc^{V}O\text{-L,L-ECD}]$	ON_2S_2	47
renal uptake	$[Tc^{III}\text{-DMSA}]$	O_xS_x	48
	$[Tc^{V}O\text{-gluceptate}]^-$	O_5	49
renal excretion	$[Tc^{IV,V}\text{-DTPA}]^-$	N_xO_x	49
	$[Tc^{V}O\text{-MAG}_3]^-$	ON_2S_2	50
myocardial perfusion	$[Tc^{I}\text{-sestamibi}]^+$	C_6	38
	$[Tc^{V}O_2\text{-tetrofosmin}]^+$	O_2P_4	51
	$[Tc^{III}\text{-furifosmin}]^+$	$N_2O_2P_2$	52
	$[Tc^{III}\text{-teboroxime}]$	N_6	53
hepatobiliary system	$[Tc^{III}\text{-HIDA}]^-$	N_2O_4	54
bone	$Tc^{IV}\text{-MDP}$	O_x	55
lung perfusion	Tc-labeled macro-aggregated albumin	–	56
blood pool imaging	Tc-labeled red blood cells	–	57
infection detection	Tc-labeled human poly-clonal immunoglobulin	–	58
tumour imaging	$[Tc^{V}O\text{-(DMSA)}_2]^-$	OS_4	48
thyroid	$Tc^{VII}O_4^-$	O_4	44

The release of ^{99}Tc to the environment due to the nuclear fuel cycle is estimated to be 1,000 TBq, while an additional discharge from the testing of nuclear weapons is estimated to be 160 TBq.[66] The estimated release of ^{99}Tc resulting from the Chernobyl accident is about 0.75 TBq.[67] The release from the nuclear fuel cycle results from reactor operations, enrichment and conversion of UF_6 in recycled nuclear fuel, and waste disposal.[65,67,68] The chemical forms of the released ^{99}Tc are thought to be TcO_4^- and volatile forms TcF_6, $HTcO_4$, TcO_3F, and Tc_2O_7.[65] $^{99}Mo/^{99m}Tc$ generator production is considered to have a minor contribution, i.e., 0.2 MBq per year per nuclear medicine department.[67]

Carpets of mosses and lichens have been used as bioindicators for the estimation of deposition of ^{99}Tc, i.e., 0.5 Bq m^{-2} at 62 °N.[69] Results from this study indicate activity concentrations of ^{99}Tc in the lichens of 0.03-0.1 Bq kg^{-1} dry weight.[67] Soil samples collected in Japan showed ^{99}Tc concentrations in the range of 0.11 to 0.46 Bq kg^{-1} dry weight.[70]

In marine and fresh water environments, TcO_4^- is rapidly taken up by wildlife, with the highest accumulations being observed for algae.[71-93] The brown alga *Fucus vesiculosus*, which has a concentration factor for Tc of 10^5 (dry weight/water), has been used, along with direct determination methods, to determine the activity concentration of ^{99}Tc in seawater.[67,71,94,95] These concentrations range from 0.03 mBq L^{-1} in the Mediterranean to 0.35 Bq L^{-1} in the Irish Sea.[67,96] As expected, areas which are exposed to releases from reprocessing facilities show an enhanced ^{99}Tc concentration.

2.6 PLANT UPTAKE OF ^{99}TC

Plants most often represent the principal means of radionuclide transfer to man due to their efficiency as scavengers of materials from air, soil, and water.[97] Information on the behaviour of radioactive materials in the biosphere is reliant on a thorough mechanistic understanding of their interaction with plants. While Cs and Sr are important because of their ability to mimic biologically essential elements, Tc, being an artificial element, is of special radioecological interest because of its extraordinary mobility and ability to concentrate in plants.

Several studies concerning the interaction of ^{99}Tc with soil and sediments have been performed.[97-111] In the absence of reducing agents, TcO_4^-, the "free" metal ion form, is very mobile in the mainly negatively charged soil environment and shows only minor interactions with inorganic solids. Several studies concerning the uptake of Tc (primarily from soil) in plants have been performed.[112-172] Results of these studies reveal that Tc in the form of TcO_4^- is rapidly taken up by plants. Data obtained by Van Loon et al.[112,113] and Hattink et al.,[114] suggest that there is a linear relationship between the Tc concentrations in spinach leaves and duckweed, respectively, and low (non-toxic) concentration of TcO_4^- in a nutrient solution. It has been suggested that the uptake mechanism for TcO_4^- is similar to that for other oxo-anions such as sulphate.[115,116] The Tc soil-to-plant transfer factors obtained do not only depend on plant species and soil type, but also on ecophysiological conditions and time (ageing effect).[117,118] This ageing effect is related to the reductive transformation of TcO_4^-, especially in association with micro-organisms[173-176] in anaerobic soil to insoluble forms such as TcO_2,[177,178] which are not taken up by plants. The uptake of other reduced Tc species, i.e., Tc-cysteine and Tc-tetren complexes, is possible, but at a much lower rate.[112,119,120]

TcO_4^- is also thought to be the main species transported within the plant.[115,118,120-124,179,180] Technetium tends to accumulate in the leaves and shoots of plants.[117,118,125] Studies with photoautotrophic organisms have indicated that TcO_4^- might be reduced to Tc^V by electrons derived from the photosynthetic transport chain. The reduction of TcO_4^- is progressively inhibited by increased concentrations of $NADP^+$, suggesting that TcO_4^- and $NADP^+$ are competing substrates of the photosynthetic electron transport chain.[126-128] Several Tc^V species appear to be formed *in vivo* by ligand exchange with complexing agents such as thiol compounds, e.g., cysteine.[112,113,119,123,126-130]

On the other hand, it was suggested,[115,131] that technetium, like selenium, can act as an analogue for sulphur in S-containing amino acids, e.g., cysteine. In higher plants, SO_4^{2-} is reduced via bound GSH reductase, followed by incorporation of reduced S^{-II} into cysteine by cysteine synthetase. These reactions are carried out in the chloroplast using reductants from the photosynthesis.[181,182] The reduction of TcO_4^- is also thought to take place in the chloroplast. However, because of the absence of the –II oxidation state, it is unlikely that Tc can act as a S-analogue in thiols. It seems more likely that the observed relation between Tc and S in these studies can be explained by the formation of a Tc-cysteine complex (see Chapter 7).

The phytotoxicity of technetium[112,132-146] is believed to result from chemical rather than radiological damage, because the calculated radiation dose from tissue-deposited Tc is not high enough to induce damage.[132,183,184] For wheat[133] and spinach grown on a Tc-containing nutrient solution,[112] toxic effects have been reported at Tc concentrations above 3.0 μmol L^{-1}. For spinach it was observed that leaves close to maturity, with little cell multiplication but intensive photosynthesis, were not very sensitive to a relative high Tc concentration. However, expanding leaves, fully in their phase of cell multiplication are much more hampered even at relatively low internal Tc concentrations.[140]

The toxic effects of technetium observed in plants are disintegration of membranes by lipid peroxidation, reduced yield, presence of necrotic areas (leaf tips), distorted interveinal areas, and other abnormal growth patterns. Lipid peroxidation,[185] which results in membrane permeability leading to cell decompartmentation (oxidative stress), might be initiated by reactive oxygen species such as superoxide anion (O_2^-), hydrogen peroxide (H_2O_2), and the hydroxyl radical (OH·). Despite the likely difference in the uptake and initial metabolism, technetium might, like other redox active metals such as copper and iron, induce membrane lipid peroxidation and oxidative stress directly by generation of reactive free radicals or indirectly, like zinc, cadmium, lead, nickel, and mercury by interaction with or binding to the antioxidant glutathione.

2.7 ^{99}TC DETERMINATION AND SPECIATION

The methods used for determination of ^{99}Tc include liquid scintillation counting (LSC),[186] inductively coupled plasma mass spectrometry (ICP-MS),[70,187-193] and, to lesser degree, neutron activation methods[194], gas flow counters,[195] graphite furnace atomic absorption spectroscopy (GF-AAS),[196] laser induced photoacoustic spectrometry (LPAS),[197] and electrochemical methods.[198] The detection limits for ^{99}Tc with LSC methods are in the range of 1-25 mBq.[63] ICP-MS might improve detection limits (2-4 mBq), although interference from ^{99}Ru, with a natural abundance of 12.7%, might be problematic, requiring chemical separation treatment. When a double focussing high-resolution system (HR-ICP-MS) is used, the detection limit is in the order of 2 μBq.[197] The most sensitive detection technique for ^{99}Tc is laser resonance ionisation mass spectrometry (RIMS).[66] The detection limit of this technique for ^{99}Tc is 0.5 μBq.[60] Activation methods for ^{99}Tc, i.e., $^{99}Tc(n,\gamma)^{100}Tc$,[194] $^{99}Tc(n,n')^{99m}Tc$,[199,200] and $^{99}Tc(\gamma,\gamma')^{99m}Tc$,[201,202] lack good precision and reliability or suffer from low sensitivity.[63] The detection limit for ^{99}Tc using a gas flow counter with an anti-coincidence shield was reported to be 1-5 mBq.[68]

A problem with the determination of ^{99}Tc in environmental materials is the lack of a stable isotopic tracer for use in determination of the chemical yield. Various radioactive isotopic tracers, such as ^{95m}Tc,[203] ^{97m}Tc,[204] and ^{99m}Tc[205] with their specific problems and limitations, have been used as yield monitors. Another approach has been the use of rhenium as a stand-in for technetium.[206] However, this approach requires full knowledge of the differences in the chemical behaviour of these two elements, e.g., redox behaviour. Generally, the isotopic yield monitors are added in the chemical form of pertechnetate. However, the assumption that the tracee (^{99}Tc) invariably exists in the same chemical form, is, at least, questionable. Therefore, if one would study the speciation of Tc in environmental materials, it is required firstly to identify the species involved, and secondly to prepare yield tracers with the same chemical form.[184] However, the multitude of (in)organic constituents in plants[208,209] makes it difficult to predict exactly what species will occur and what artefacts might be encountered.[210-213]

Conventional speciation methods rely on the combination of a separation technique, e.g., liquid chromatography,[214] and a selective determination technique, e.g., radiometric methods or mass spectrometry. Another more direct type of speciation study for ^{99}Tc might consist of either *in situ* spectroscopic techniques,[215] such as *in vivo* nuclear magnetic resonance (NMR) and electron paramagnetic resonance (EPR) spectroscopy, or *in vivo* chemical sensor techniques. However, these direct methods generally suffer from experimental difficulties. A correlation chart of ^{99}Tc (nuclear spin: 9/2) NMR chemical shifts and oxidation states of Tc has been published,[216] but this technique[217-220] suffers from a low sensitivity.[221] EPR techniques[222] are not very useful in environmental study, since only the spectra of not very relevant oxidation states

(Tc^0, Tc^{II}, and Tc^{VI}) can be detected at convenient temperatures. Until now, it was only possible to develop chemical sensors (both a voltammetric electrochemical and a fiber optic absorbance sensor) for *in vivo* monitoring of Re species, but not for analogous Tc species.[43]

2.8 REFERENCES

(1) Mendeleev, D. I. *J. Russ. Chem. Soc.* **1869**, *1*, 60.
(2) De Jonge, F. A. A.; Pauwels, E. K. J. *Eur. J. Nucl. Med.* **1996**, *23*, 336-344.
(3) Kuroda, P. K. *Radiochim. Acta* **1993**, *63*, 9-18.
(4) Yoshihara, H. K. *Radiochim. Acta* **1997**, *77*, 9-13.
(5) Ogawa, M. *Chem. News* **1908**, *98*, 249.
(6) Noddack, W.; Tacke, I.; Berg, O. *Naturwissenschaften* **1925**, *13*, 567-574.
(7) Boyd, G. E. *J. Chem. Edu.* **1959**, *36*, 3-14.
(8) Perrier, C.; Segré, E. *Nature* **1937**, *140*, 193-194.
(9) Perrier, C.; Segré, E. *J. Chem. Phys.* **1937**, *5*, 712-716.
(10) Seaborg, G. T.; Segré, E. *Phys. Rev.* **1939**, *55*, 808-814.
(11) Segré, E.; Wu, C. S. *Phys. Rev.* **1940**, *57*, 552.
(12) Lincoln, D. C.; Sullivan, W. H. in *National Nuclear Energy Series-IV*, Vol. 9; McGraw-Hill: New York, 1951; p 778.
(13) Sullivan, R. P. in *National Nuclear Energy Series-IV*, Vol. 9; McGraw-Hill: New York, 1951; p 783.
(14) Motta, E. E.; Boyd, G. E.; Larson, Q. V. *Phys. Rev.* **1947**, *72*, 1270.
(15) Committee on New Elements, IUPAC, *Chem. Engr. News* **1946**, *27*, 2996.
(16) Firestone, R. B.; Shirley, V. S.; Chu, S. Y. F.; Baglin, C. M.; Zipkin, J. *Table of isotopes*, 8th ed., CD-ROM v. 1.0; Wiley: New York, 1996.
(17) Deutsch, E.; Libson, K.; Vanderheyden, J.-L.; Ketring, A. R.; Maxon, H. R. *Nucl. Med. Biol.* **1986**, *13*, 465.
(18) Schwochau, K.; Pleger, U. *Radiochim. Acta* **1993**, *63*, 103-110.
(19) Baldas, J. *Adv. Inorg. Chem.* **1994**, *41*, 1-123.
(20) Johannsen, B.; Spies, H. *Isotopenpraxis* **1988**, *24*, 449-454.

(21) Schwochau, K. *Radiochim. Acta* **1983**, *32*, 139.
(22) Deutsch, E.; Libson, K.; Jurisson, S.; Lindoy, L. F. *Prog. Inorg. Chem.* **1983**, *30*, 75-139.
(23) Melnik, M.; Van Lier, J. E. *Coord. Chem. Rev.* **1987**, *77*, 275-324.
(24) Clarke, M. J.; Podbielski, L. *Coord. Chem. Rev.* **1987**, *78*, 253-331.
(25) Turp, N.; Turp, J. E. *Coord. Chem. Rev.* **1987**, *80*, 157-171.
(26) Constable, E. C.; Housecroft, C. E. *Coord. Chem. Rev.* **1994**, *131*, 153-175.
(27) Housecroft, C. E. *Coord. Chem. Rev.* **1995**, *142*, 21-41.
(28) Housecroft, C. E. *Coord. Chem. Rev.* **1997**, *162*, 305-318.
(29) *Technetium in Chemistry and Nuclear Medicine*, Deutsch, E., Nicolini, M., Wagner, Jr., H. N. Eds.; Cortina International: Verona, 1983.
(30) *Technetium in Chemistry and Nuclear Medicine 2*, Nicolini, M., Bandoli, G., Mazzi, U., Eds.; Cortina International: Verona, 1986.
(31) *Technetium and Rhenium in Chemistry and Nuclear Medicine 3*, Nicolini, M., Bandoli, G., Mazzi, U., Eds.; Cortina International: Verona, 1990.
(32) *Technetium and Rhenium in Chemistry and Nuclear Medicine 4*, Nicolini, M., Bandoli, G., Mazzi, U., Eds.; Sgeditoriali: Padova, 1995.
(33) Davison, A.; Jones, A. G. *Int. J. Appl. Radiat. Isot.* **1982**, *33*, 875-881.
(34) Baldas, J. in *Topics in Current Chemistry 176: Technetium and Rhenium*; Yoshihara, K., Omori, T., Eds.; Springer: Berlin, 1996; pp 37-76.
(35) Johannsen, B.; Spies, H. in *Topics in Current Chemistry 176: Technetium and Rhenium*; Yoshihara, K., Omori, T., Eds.; Springer: Berlin, 1996; pp 77-121.
(36) Omori, T. in *Topics in Current Chemistry 176: Technetium and Rhenium*; Yoshihara, K., Omori, T., Eds.; Springer: Berlin, 1996; pp 253-273.
(37) Kryutchkov, S. V. in *Topics in Current Chemistry 176: Technetium and Rhenium*; Yoshihara, K., Omori, T., Eds.; Springer: Berlin, 1996; pp 189-252.
(38) Holman, B. L.; Sporn, V.; Jones, A. G.; Sia, S. T. B. *J. Nucl. Med.* **1987**, *28*, 13.
(39) Bailey, M. F.; Dahl, L. F.; *Inorg. Chem.* **1965**, *4*, 1140.
(40) Alberto, R. in *Topics in Current Chemistry 176: Technetium and*

Rhenium; Yoshihara, K., Omori, T., Eds.; Springer: Berlin, 1996; pp 149-187.
(41) Hileman, J. C.; Huggins, D. K.; Kaesz, H. D. *Inorg. Chem.* **1962**, *1*, 933.
(42) Heineman, W. R.; Swaile, B. H.; Blubaugh, E. A.; Landis, D. A.; Seliskar, C. J.; Deutsch, E. *Radiochim. Acta* **1993**, *63*, 199-203.
(43) Molinski, V. J. *Int. J. Appl. Radiat. Isot.* **1982**, *33*, 811-819.
(44) Jurisson, S.; Berning, D.; Wei Jia; Dangshe Ma *Chem. Rev.* **1993**, *93*, 1137-1156.
(45) Deutsch, E. *Radiochim. Acta* **1993**, *63*, 195-197.
(46) Neirinckx, R. D.; Canning L.; Piper I. M.; Nowotnik, D. P.; Pickett, R. D.; Holmes, R. A.; Volkert, W. A.; Forster, A. M.; Weisner, P. S.; Marriott, J. A.; Chaplin, S. B. *J. Nucl. Med.* **1987**, *28*, 191.
(47) Walovitch, R. C.; Hill T. C.; Garrity S. T.; Cheesman, E. H.; Burgess, B. A.; O'Leary, D. A.; Watson, A. D.; Ganey, M. V.; Morgan, R. A.; Williams, S. J. *J. Nucl. Med.* **1989**, *30*, 1892-1901.
(48) Blower, P. J.; Singh, J.; Clarke, S. E. M. *J. Nucl. Med.* **1991**, *32*, 845.
(49) Steigman, J.; Eckelman, W. C. *The chemistry of technetium in medicine*; National Academy Press: Washington DC, 1992.
(50) Fritzberg, A. R.; Kasina, S.; Eshima, D.; Johnson, D. L. *J. Nucl. Med.* **1986**, *27*, 111.
(51) Kelly, J. D.; Forster, A. M.; Higley, B.; Archer, C. M.; Booker, F. S.; Canning, L. R.; Chiu, F. W.; Edwards, B.; Gill, H. K.; McPartlin, M.; Nagle K. R.; Latham, I. A.; Pickett, R. D.; Storey, A. E.; Webbon, P. M. *J. Nucl. Med.* **1993**, *34*, 222-227.
(52) DeRosch, M. A.; Brodack, J. W.; Grunmon, G. D.; Marmion, M.; Nosco, D. L.; Deutsch, K. F.; Deutsch, E. *J. Nucl. Med.* **1992**, *33*, 850.
(53) Jurisson, S.; Francesconi, L. F.; Linder, K. E.; Treher, E.; Malley, M. F.; Gougoutas, J. Z.; Nunn, A. D. *Inorg. Chem.* **1991**, *30*, 1820.
(54) Loberg, M. D.; Fields, A. T. *Int. J. Appl. Rad. Isot.* **1978**, *29*, 167.
(55) Libson, K.; Deutsch, E.; Barnett, B. L. *J. Am. Chem. Soc.* **1980**, *102*, 2476.
(56) Robbins, P. J.; Feller, P. A.; Nishiyama, H. *Health Phys.* **1976**, *30*, 173-178.
(57) Neumann, P.; Schicha, H.; Schürnbrand, P.; Bähre, M.; Emrich, D. *Eur. J. Nucl. Med.* **1983**, *8*, 463-466.

(58) Buscombe, J. R.; Lui, D.; Ell, P. J. *Eur. J. Nucl. Med.* **1990**, *16*, 649-655.
(59) Hidaka, H.; Shinotsuka, K.; Hollinger, Ph. *Radiochim. Acta* **1993**, *63*, 19-22.
(60) Kenna, B. T.; Kuroda, P. K. *J. Inorg. Nucl. Chem.* **1961**, *23*, 142-144.
(61) Schulte, E. H.; Scoppa, P. *Sci. Total Environ.* **1987**, *64*, 163-179.
(62) Ehrhardt, K. C.; Attrep, Jr., M. *Environ. Sci. Technol.* **1978**, *12*, 55-57.
(63) Yoshihara, K. in *Topics in Current Chemistry 176: Technetium and Rhenium*; Yoshihara, K., Omori, T., Eds.; Springer: Berlin, 1996; pp 17-35.
(64) Katcoff, S. *Nucleonics* **1958**, *16*, 78.
(65) Till, J. E. in *Technetium in the environment*; Desmet, G. M., Myttenaere, C., Eds.; Elsevier: London, 1986; pp 1-20.
(66) Trautmann, N. *Radiochim. Acta* **1993**, *63*, 37-43.
(67) Holm, E. *Radiochim. Acta* **1993**, *63*, 57-62.
(68) Till, J. E.; Hoffman, F. O.; Dunning, Jr., D. E. *Health Phys.* **1979**, *36*, 21-30.
(69) Holm, E.; Rioseco, J. *J. Environ. Radioact.* **1987**, *5*, 343.
(70) Tagami, K.; Uchida, S. *Radiochim. Acta* **1993**, *63*, 69-72.
(71) Hirano, S.; Matsuba, M. *Radiochim. Acta* **1993**, *63*, 79-82.
(72) Aarkrog, A. *Estuarine, Coastal and Shelf Science* **1986**, *24*, 637.
(73) Beasley, T. M.; Lorz, H. V. in *Technetium in the environment*; Desmet, G. M., Myttenaere, C., Eds.; Elsevier: London, 1986; pp 197-216.
(74) Benco, C.; Cannarsa, S.; Ceppodomo, I.; Zattera, A. in *Technetium in the environment*; Desmet, G. M., Myttenaere, C., Eds.; Elsevier: London, 1986; pp 217-227.
(75) Blaylock, B. G.; Frank, M. L.; Hoffman, F. O.; DeAngelis, D. L. in *Technetium in the environment*; Desmet, G. M., Myttenaere, C., Eds.; Elsevier: London, 1986; pp 79-90.
(76) Garcia-Leon, M.; Piazza, C.; Madurga, G. in *Technetium in the environment*; Desmet, G. M., Myttenaere, C., Eds.; Elsevier: London, 1986; pp 169-177.
(77) Holm, E.; Rioseco, J.; Mattson, S. in *Technetium in the environment*; Desmet, G. M., Myttenaere, C., Eds.; Elsevier: London, 1986; pp 61-68.
(78) Bonotto, S. *Mar. Pollut. Bull.* **1988**, *19*, 61.
(79) Aarkrog, A.; Dahlgaard, H.; Hallstadius, L.; Holm, E.; Mattson, S.;

Rioseco, J. in *Technetium in the environment*; Desmet, G. M., Myttenaere, C., Eds.; Elsevier: London, 1986; pp 69-78.

(80) Gromov, V. V. *Mar. Sci. Comm.* **1976**, *2*, 227-247.

(81) Fisher, N. S. *Environ. Sci. Technol.* **1982**, *16*, 579-581.

(82) Beasley, T. M.; Gonor, J. J.; Lorz, H. V. *Mar. Environ. Res.* **1982**, *7*, 103-116.

(83) Jeanmaire, L.; Masson, M.; Patti, F.; Germain, P.; Cappellini, L. *Mar. Pollut. Bull.* **1981**, *12*, 29-32.

(84) Topcuoglu, S.; Fowler, S. W. *Mar. Environ. Res.* **1984**, *12*, 25-43.

(85) Bonotto, S.; Kirchmann, R.; Van Baelen, J.; Hurtgen, C.; Cogneau, M.; Van der Ben, D.; Verthe, C.; Bouquegneau, J. M. in *Speciation of fission and activation products in the environment*; Bulman, R. A., Bulman, J. R., Eds.; Elsevier: London, 1986; pp 382-390.

(86) Van der Ben, D.; Cogneau, M.; Robbrecht, V.; Nuyts, G.; Bossus, A.; Hurtgen, C.; Bonotto, S. *Mar. Pollut. Bull.* **1990**, *21*, 84-86.

(87) Spies, R. B. *Health Phys.* **1975**, *29*, 695-699.

(88) Beasley, T. M.; Lorz, H. V.; Gonor, J. J. *Health Phys.* **1982**, *43*, 501-507.

(89) Verthe, C.; Moreau, Z.; Mania, B.; Van Baelen, J.; Van der Ben, D.; Cogneau, M.; Vandecasteele, C. M.; Bouquegneau, J. M.; Myttenaere, C.; Bonotto, S. in *Technetium in the environment*; Desmet, G. M., Myttenaere, C., Eds.; Elsevier: London, 1986; pp 319-339.

(90) Garnham, G. W.; Codd, G. A.; Gadd, G. M. *Appl. Microbiol. Biotechnol.* **1992**, *37*, 679-684.

(91) Hurtgen, C.; Koch, G.; Van der Ben, D.; Bonotto, S. *Sci. Total Environ.* **1988**, *70*, 131-149.

(92) Blaylock, B. G.; Frank, M. L.; DeAngelis, D. L. *Health Phys.* **1982**, *42*, 257-266.

(93) Zeevaert, Th.; Vandecasteele, C. M.; Kirchmann, R. *Health Phys.* **1989**, *57*, 337-343.

(94) Roos, P.; Holm, E.; Sanchez, A.; Gastaud, J.; Noshkin, V.; Ballestra, S. *Sci. Total Environ.* **1993**, *130-131*, 1-22.

(95) Riley, J. P.; Siddiqui, S. A. *Mar. Pollut. Bull.* **1986**, *17*, 229.

(96) Leonard, K. S.; McCubbin, D.; Brown, J.; Bonfield, R.; Brooks, T. *Mar. Pollut. Bull.* **1997**, *34*, 628-636.

(97) Shaw, G.; Bell, J. N. B. in *Plants and the chemical elements*; Farago,

M. E., Ed.; VCH: Weinheim, 1994; 179-220.
(98) Gast, R. G. Rapport US AE no. At (11-1)-2447, 1975, 17-22.
(99) Wildung, R. E.; McFadden, K. M.; Garland, T. R. *J. Environ. Qual.* **1979**, *8*, 156-161.
(100) Hoffman, F. O. *Environ. Sci. Technol.* **1982**, *16*, 214.
(101) Stalmans, M.; Maes, A.; Cremers, A. in *Technetium in the environment*; Desmet, G. M., Myttenaere, C., Eds.; Elsevier: London, 1986; pp 91-113.
(102) Van Loon, L. R.; Stalmans, M.; Maes, A.; Cremers, A.; Cogneau, M. in *Technetium in the environment*; Desmet, G. M., Myttenaere, C., Eds.; Elsevier: London, 1986; pp 143-153.
(103) Wildung, R. E.; Garland, T. R.; McFadden, K. M.; Cowan, C. E. in *Technetium in the environment*; Desmet, G. M., Myttenaere, C., Eds.; Elsevier: London, 1986; pp 115-129.
(104) Sheppard, M. I.; Evenden, W. G. *J. Environ. Qual.* **1985**, *14*, 554.
(105) Vandecasteele, C. M.; Dehut, J. P.; Van Laer, S.; Deprins, D.; Myttenaere, C. *Health Phys.* **1989**, *57*, 247-254.
(106) Fowler, S. W.; Aston, S. R.; Benayoun, G.; Parsi, P. *Mar. Environ. Res.* **1983**, *8*, 87-100.
(107) Sheppard, M. I.; Thibault, D. H. *J. Environ. Qual.* **1991**, *20*, 101-114.
(108) Del Cul, G. D.; Bostick, W. D.; Trotter, D. R.; Osborne, P. E. *Sep. Sci. Technol.* **1993**, *28*, 551-564.
(109) Lieser, K. H.; Bauscher, Ch. *Radiochim. Acta* **1987**, *42*, 205-213.
(110) Sekine, T.; Watanabe, A.; Yoshihara, K.; Kim, J. I. *Radiochim. Acta* **1993**, *63*, 87-90.
(111) Masson, M.; Patti, F.; Colle, C.; Roucoux, P.; Grauby, A.; Saas, A. *Health Phys.* **1989**, *57*, 269-279.
(112) Van Loon, L. R. *PhD thesis*; K. U. Leuven, 1986.
(113) Van Loon, L. R.; Desmet, G. M.; Cremers, A. *Health Phys.* **1989**, *57*, 309-314.
(114) Hattink, J.; Harms, A. V.; Weltje, L.; Verburg, T. G.; Wolterbeek, H. Th. **1998**, in preparation.
(115) Cataldo, D. A.; Garland, T. R.; Wildung, R. E. in *Technetium in the environment*; Desmet, G. M., Myttenaere, C., Eds.; Elsevier: London, 1986; pp 265-280.
(116) Cataldo, D. A.; Wildung, R. E.; Garland, T. R. *Plant Physiol.* **1983**, *73*,

849-852.
(117) Mousny, J. M.; Myttenaere, C. *Plant Soil* **1981**, *61*, 403-412.
(118) Mousny, J. M.; Roucoux, P.; Myttenaere, C. *Environ. Exp. Bot.* **1979**, *19*, 263-268.
(119) Van Loon, L. R.; Lembrechts, J. F. M. M. in *Technetium in the environment*; Desmet, G. M., Myttenaere, C., Eds.; Elsevier: London, 1986; pp 301-306.
(120) Van Loon, L. R.; Desmet, G. M.; Cremers, A. in *Speciation of fission and activation products in the environment*; Bulman, R. A., Cooper, J. R., Eds.; Elsevier: London, 1986; pp 352-360.
(121) Cataldo, D. A.; Wildung, R. E.; Garland, T. R. in *Environmental chemistry and cycling processes*, proceedings of a DOE symposium No. 45 Washington. Conf.-760429, 1978; pp 537-549.
(122) Desmet, G. M. in *Technetium in the environment*; Desmet, G. M., Myttenaere, C., Eds.; Elsevier: London, 1986; pp 189-195.
(123) Lembrechts, J. F. M. M.; Van Loon, L. R. in *Technetium in the environment*; Desmet, G. M., Myttenaere, C., Eds.; Elsevier: London, 1986; pp 295-300.
(124) Cataldo, D. A.; McFadden, K. M.; Garland, T. R.; Wildung, R. E. in *Speciation of fission and activation products in the environment*; Bulman, R. A., Cooper, J. R., Eds.; Elsevier: London, 1986; pp 398-408.
(125) Myttenaere, C.; Vandecasteele, C. M.; Roucoux, P.; Lietart, A.; Mousny, J. M. in *Technetium in the environment*; Desmet, G. M., Myttenaere, C., Eds.; Elsevier: London, 1986; pp 281-293.
(126) Lembrechts, J. F. *Ph. D. thesis*; U. I. Antwerpen, 1986.
(127) Lembrechts J. F.; Desmet, G. M. *Plant Physiol.* **1986**, *81*, 1003-1007.
(128) Lembrechts J. F.; Desmet, G. *Health Phys.* **1989**, *57*, 255-262.
(129) Lembrechts, J. F.; Desmet, G. M.; Overbeek, H. *Environ. Exp. Bot.* **1985**, *25*, 355-360.
(130) Lembrechts J. F.; Desmet, G. M. *Environ. Exp. Bot.* **1985**, *25* 361-368.
(131) Cataldo, D. A.; Garland, T. R.; Wildung, R. E.; Fellows, R. J. *Health Phys.* **1989**, *57*, 281-287.
(132) Sheppard, M. I.; Vandergraaf, T. T.; Thibault, D. H.; Keith Reid, J. A. *Health Phys.* **1983**, *44*, 635-643.

(133) Landa, E. R.; Hart Thorvig, L. J.; Gast, R. G. in *Biological implications of metals in the environment* Druckler, H.; Wildung, R. E., Eds., USDERA SYMP SER Conf.- 750959 NTIS, Springfield, 1977, pp 390-401.
(134) Wildung, R. E.; Garland, T. R.; Cataldo, D. A. *Health Phys.* **1977**, *32*, 314-317.
(135) Gearing, P.; Van Baalen, C.; Parker, P. L. *Radionuclides in the environment*, proceedings of the 3rd symposium Radioecology. CONF-710501-Pz., 1975; pp 857-865.
(136) Gearing, P.; Van Baalen, C.; Parker, P. L. *Plant Physiol.* **1975**, *55*, 240-246.
(137) Bennassar, A.; Vazquez, M. D.; Cabot, C.; Poschenrieder, Ch.; Barcelo, J. *Water, Air, Soil Pollut.* **1991**, *57-58*, 681-689.
(138) Berlyn, G. P.; Dhillon, S. S.; Koslow, E. E. *Environ. Manage.* **1980**, *4*, 149-156.
(139) Woodard-Blankenship, B.; Neel, J. W.; Papin, P. J. *Water, Air, Soil Pollut.* **1995**, *81*, 411-428.
(140) Desmet, G. M.; Dirkse, W. G. *Environ. Exp. Bot.* **1984**, *24*, 45-52.
(141) Garten, C. T.; Lomax, R. D. *Health Phys.* **1989**, *57*, 299-307.
(142) Neel, J. W.; Onasch, M. A. *Health Phys.* **1989**, *57*, 289-298.
(143) Vazquez, M. D.; Bennassar, A.; Cabot, C.; Poschenrieder, Ch.; Barcelo, J. *Environ. Exp. Bot.* **1990**, *30*, 271.
(144) Onasch, M. A. *Master's thesis*; San Diego State University, San Diego, 1984; pp 42-51.
(145) Fujimoto, *Annual report of the radiation center of Osaka prefecture* **1985**, *26*, 123.
(146) Fujimoto, *Annual report of the radiation center of Osaka prefecture* **1987**, *28*, 67.
(147) Degenkolb, S. J.; Neel, J. W.; Papin, P. J. *J. Environ. Qual.* **1994**, *23*, 1265-1271.
(148) Echevarria, G.; Vong, P. C.; Leclerc-Cessac, E.; Morel, J. L. *J. Environ. Qual.* **1997**, *26*, 947-956.
(149) Sparkes, S. T.; Long, S. E. *Report AERE-R 12743 DOE/RW/88098*; Harwell Laboratory Oxfordshire, 1988.
(150) Harms, A. V.; Van Elteren, J. T.; Wolterbeek, H. Th.; De Goeij, J. J. M.

1998, in preparation.
(151) Green, N.; Wilkins, B. T. *Sci. Total Environ.* **1995**, *173-174*, 385-388.
(152) Harms, A. V.; Van Elteren, J. T. *J. Radioanal. Nucl. Chem.* **1998**, *227*, in press.
(153) Harms, A. V.; Van Elteren, J. T. **1998**, in preparation.
(154) Myttenaere, C.; Daoust, C.; Roucoux, P. *Environ. Exp. Bot.* **1980**, *20*, 415-419.
(155) Hoffman, F. O.; Garten, Jr., C. T.; Huckabee, J. W.; Lucas, D. M. *J. Environ. Qual.* **1982**, *11*, 134-141.
(156) Garten, Jr., C. T. *Environ. Intern.* **1987**, *13*, 311-321.
(157) Garten, Jr., C. T.; Tucker, C. S.; Walton, B. T. *J. Environ. Radioactivity* **1986**, *3*, 163-188.
(158) Garten, Jr., C. T.; Tucker, C. S.; Scott, T. G. *J. Environ. Radioactivity* **1986**, *4*, 91-99.
(159) Yanagisawa, K.; Muramatsu, Y. *Radiochim. Acta* **1993**, *63*, 83-86.
(160) Murphy, Jr., C. E.; Johnson, T. L. *J. Environ. Qual.* **1993**, *22*, 793-797.
(161) Gerber, G. B.; Van Hees, M.; Garten, Jr., C. T.; Vandecasteele, C. M.; Vankerkom, J.; Van Bruwaene, R.; Kirchmann, R.; Colard, J.; Cogneau, M. *Health Phys.* **1989**, *57*, 315-319.
(162) Müller, H.; Ter Meer-Bekk, C. H. *J. Radioanal. Nucl. Chem. Art.* **1986**, 181-185.
(163) Echevarria, G.; Vong, P. C.; Morel, J. L. *J. Environ. Radioact.* **1998**, *38*, 163-171.
(164) Yanagisawa, K.; Muramatsu, Y.; Kamada, H. *Radioisotopes* **1992**, *41*, 397-402.
(165) Desmet, G. M.; Van Loon, L. R.; Howard, B. J. *Sci. Total Environ.* **1991**, *100*, 105-124.
(166) Garten, C. T.; Myttenaere, C.; Vandecasteele, C. M.; Kirchmann, R.; Van Bruwaene, R. in *Technetium in the environment*; Desmet, G. M., Myttenaere, C., Eds.; Elsevier: London, 1986; pp 319-332.
(167) Scopinaro, F.; De Vincentis, G.; Pani, R.; Pellegrini, R.; Banci, M.; Casu, C.; Lerardi, M.; Russo, A.; Soluri, A. *Nucl. Med. Commun.* **1994**, *15*, 905-915.
(168) Finch, R. R. (1983) *Master's thesis*. San Diego State University, San Diego, 1983, 38-49.

(169) Dehut, J. P.; Fonsny, K.; Myttenaere, C.; Van Laer, S.; Deprins, D.; Vandecasteele, C. M. *Health Phys.* **1989**, *57*, 263-267.
(170) Vandecasteele, C. M.; Garten, Jr., C. T.; Van Bruwaene, R.; Janssens, J.; Kirchmann, R.; Myttenaere, C. in *Speciation of fission and activation products in the environment*; Bulman, R. A., Cooper, J. R., Eds.; Elsevier: London, 1986; pp 368-380.
(171) Cataldo, D. A.; Wildung, R. E. *Sci. Total Environ.* **1983**, *28*, 159-168.
(172) Gast, R. G.; Hart Thorvig, L.; Landa, E. R.; Gallagher, K. J. in *Biological implications of metals in the environment*; Druckler, H., Wildung, R. E., Eds.; USDER SYMP SER Conf.- 750959 NTIS, Springfield, 1977; pp 550-567.
(173) Henrot, J. *Health Phys.* **1989**, *57*, 239-245.
(174) Tagami, K.; Uchida, S. *Chemosphere* **1996**, *33*, 217-225.
(175) Mania, B.; Moreau, Z.; Van der Ben, D.; Van Baelen, J.; Verthé, C.; Bouquegneau, J. M.; Cogneau, M.; Bonotto, S.; Vandecasteele, C. M.; Pignolet, L.; Myttenaere, C.; in *Technetium in the environment*; Desmet, G. M., Myttenaere, C., Eds.; Elsevier: London, 1986; pp 229-244.
(176) Lloyd, J. R.; Cole, J. A.; Macaskie, L. E. *J. Bacteriol.* **1997**, *179*, 2014-2021.
(177) Meyer, R. E.; Arnold, W. D.; Case, F. I.; O'Kelley, G. D. *Radiochim. Acta* **1991**, *55*, 11-18.
(178) Burnett, K. B.; Campbell, A. B.; Jobe, D. J.; Lemire, R. J.; Taylor, P. *Radiochim. Acta* **1995**, *69*, 241-249.
(179) Krijger, G. C.; Claessens, H. A.; Wolterbeek, H. Th. *Chem. Spec. Bioavail.* **1996**, *8*, 29-36.
(180) Krijger, G. C.; Van Vliet, P. M.; Wolterbeek, H. T. *J. Exp. Bot.* **1998**, in preparation.
(181) Schwenn, J. D. *Z. Naturforsch.* **1994**, *49c*, 531-539.
(182) Leustek, Th. *Physiol. Plant.* **1996**, *97*, 411-419.
(183) Gerber, G. B.; Hegela, M.; Vankerkom, J.; Kirchmann, R.; Maison, J. R.; Lambiet-Collier, M. *Health Phys.* **1989**, *57*, 345-350.
(184) Coffey, J. L.; Hayes, R. L.; Rafter, J. J.; Watson, E. E.; Carlton, J. E.; *Health Phys.* **1984**, *46*, 418-422.
(185) Vangronsveld, J.; Clijsters, H. in *Plants and the chemical elements*; Farago, M. E., Ed.; VCH: Weinheim, 1994; 149-177.

(186) Walker, C. R.; Short, B. W.; Spring, H. S. in *Radioelement Analysis; Progress and problems*; Lyon, W. S., Ed.; Ann Arbor Science: Ann Arbor, 1980; pp 101-110.
(187) Morita, S.; Kim, C. K.; Takaku, Y.; Seki, R.; Ikeda, N. *Appl. Radiat. Isot.* **1991**, *42*, 531.
(188) Morita, S.; Tobita, K.; Kurabayashi, M. *Radiochim. Acta* **1993**, *63*, 63-67.
(189) Momoshima, N.; Sayad, M.; Takashima, Y. *Radiochim. Acta* **1993**, *63*, 73-78.
(190) Tagami, K.; Uchida, S. *Appl. Radiat. Isot.* **1996**, *47*, 1057-1060.
(191) Richter, R. C.; Koirtyohann, S. R.; Jurisson, S. S. *J. Anal. Atom. Spectrom.* **1997**, *12*, 557-562.
(192) Dixon, P.; Curtis, D. B.; Musgrave, J.; Roensch, F.; Roach, J.; Rokop, D. *Anal. Chem.* **1997**, *69*, 1692-1699.
(193) Nicholson, S.; Sanders, T. W.; Blaine, L. M. *Sci. Total Environ.* **1993**, *130-131*, 275-284.
(194) Foti, S.; Delucchi, E.; Akamian, V. *Anal. Chim. Acta* **1972**, *60*, 269-276.
(195) Holm, E.; Rioseco, J.; Garcia-Leon, M. *Nucl. Instr. Meth. Phys. Res.* **1984**, *223*, 204.
(196) Kaye, J. H.; Ballou, N. E. *Anal. Chem.* **1978**, *50*, 2076-2078.
(197) Fujita, T.; Sekine, T.; Hiraga, H.; Yoshihara, K.; Mutalib, A.; Alberto, R.; Kim, J. I. *Radiochim. Acta* **1993**, *63*, 45-47.
(198) Friedrich, M.; Ruf, H. *J. Electroanal. Chem.* **1986**, *198*, 261-268.
(199) Ikeda, N.; Seki, R.; Kamemoto, M.; Otsuji, M. *J. Radioanal. Nucl. Chem.* **1989**, *131*, 65.
(200) Yamagishi, I.; Kubota, M.; Sekine, T.; Yoshihara, K. *Radiochim. Acta* **1993**, *63*, 33-36.
(201) Sekine, T.; Yoshihara, K.; Németh, Zs.; Lakosi, L.; Veres, Á *J. Radioanal. Nucl. Chem.* **1989**, *130*, 269.
(202) Lakosi, L.; Sáfár, J.; Veres, Á.; Sekine, T.; Yoshihara, K. *Radiochim. Acta* **1993**, *63*, 23-28.
(203) Chu, N. Y.; Feldstein, J. *Talanta* **1984**, *31*, 809.
(204) Kaye, J. H.; Merill, J. A.; Kinnison, R. R.; Rapids, M. S.; Ballou, N. E. *Anal. Chem.* **1982**, *54*, 1158-1163.
(205) Martinez-Lobo, A.; Garcia-Leon, M.; Madurga, G. *Appl. Radiat Isot.*

1986, *37*, 438-440.

(206) Harvey, B. R.; Williams, K. J.; Lovett, M. B.; Ibbett, R. D. *J. Radioanal. Nucl. Chem.* **1992**, *158*, 417-436.

(207) Harvey, B. R.; Lovett, M. B. *Nucl. Instr. Meth. Phys. Res.* **1984**, *223*, 224-234.

(208) Mehra, A.; Farago, M. E. in *Plants and the chemical elements*; Farago, M. E., Ed.; VCH: Weinheim, 1994; 31-66.

(209) Robinson, T. *The organic constituents of higher plants*, 4th Ed.; Cordus Press: North Amherst, 1980.

(210) Harms, A. V.; Wolterbeek, H. Th.; Woroniecka, U. D. *Appl. Radiat. Isot.* **1996**, *47*, 605-610.

(211) Harms, A. V.; Van Elteren J. T.; Claessens, H. A. *J. Chromatogr. A* **1996**, *755*, 219-225.

(212) Homer, F. A.; Reeves, R. D.; Brooks, R. R.; Baker, A. J. M. *Phytochemistry* **1991**, *30*, 2141-2145.

(213) Farago, M. E.; Mahmoud, I. E. D. A. W. *Min. Environ.* **1983**, *5*, 113-121.

(214) Chau, Y. K. *Analyst* **1992**, *117*, 571-575.

(215) Wolf, W. R. in *The Importance of Chemical "Speciation" in Environmental Processes*; Springer-Verlag: Berlin, 1986; pp 39-58.

(216) O'Connell, L. A.; Pearlstein, R. M.; Davison, A.; Thornback, J. R.; Kronauge, J. F.; Jones, A. G. *Inorg. Chim. Acta* **1989**, *161*, 39-43.

(217) Davison, A.; Kronauge, J. F.; Jones, A. G.; Pearlstein, R. M.; Thornback, J. R. *Inorg. Chem.* **1988**, *27*, 3245.

(218) Davison, A.; Kronauge, J. F.; O'Connell, L. A.; Pearlstein, R. M.; Thornback, J. R.; Jones, A. G. *J. Labelled Compd. Radiopharm.* **1989**, *26*, 32.

(219) Kronauge, J. F.; Piwnica-Worms, D.; Davison, A.; LeFurgey, A.; Ingram, P.; Lieberman, M.; Holman, B. L.; Jones, A. G. *J. Labelled Compd. Radiopharm.* **1991**, *30*, 62A.

(220) Piwnica-Worms, D.; Kronauge, J. F.; LeFurgey, A.; Backus, M.; Hockett, D.; Ingram, P.; Lieberman, M.; Holman, B. L.; Jones, A. G.; Davison, A. *Mag. Res. Imaging* **1994**, *12*, 641.

(221) Van Vliet, P. M.; Krijger, G. C. unpublished results.

(222) Abram, U.; Kirmse, R. *Radiochim. Acta* **1993**, *63*, 139-143.

Chapter 3

The Use of 2-Mercaptoethanol in Plant Technetium Speciation Procedures*

3.1 INTRODUCTION

^{99}Tc, a β^{-}-emitter ($E_{\beta^{-},max}$ = 292 keV) with a long half-life ($t_{½}$ = 2.13×10^{5} y), is produced by fission of nuclear fuel (isobaric fission yield of ^{235}U for A = 99 is 6.1%). The main sources of technetium to the environment are the nuclear power generation and associated industry and nuclear weapon testing programs.[1,2] The dominant chemical form of technetium in aerobic environments is pertechnetate.[3] Because of the high solubility and mobility of pertechnetate in soil and the large soil-to-plant transfer factor of technetium,[4,5] a good understanding of the uptake, the distribution and the metabolism of technetium species in plants is necessary. The available data suggest an uptake of pertechnetate and a metabolic transformation of it in plants to species of reduced technetium.[6-10] A large amount of these data is based on experiments in which 2-mercaptoethanol is added to the homogenisation buffer of plant material or to the mobile phase of a chromatographic separation of the technetium species. Isolation procedures for plant proteins often include a reductor like 2-mercaptoethanol in the homogenisation buffer in order to inhibit the activity of polyphenol oxidase.[11,12] Reaction products of polyphenol oxidase

---

* This chapter is a slightly modified version of the paper "A Caveat on the Use of 2-Mercaptoethanol in Plant Technetium Speciation Procedures"; Harms, A. V.; Wolterbeek, H. Th.; Woroniecka, U. D. *Appl. Radiat. Isot.* **1996**, *47*, 605-610.

may react with the thiol and amino groups of proteins and, consequently, may inhibit the isolation.

However, binding of ^{99m}Tc to proteins can be achieved after reduction of the disulphide groups with 2-mercaptoethanol. This results in very stable Tc-protein compounds.[13,14] Stable Tc-plant-protein may unintentionally be formed in the same way, when using 2-mercaptoethanol in plant technetium speciation procedures. 2-Mercaptoethanol, which is apart from a reducing compound also ligand for $[Tc^{V}O]^{3+}$,[15] may also react with both pertechnetate and relatively labile plant-formed Tc species.

Working with the anionic Tc-DTPA complex,[16-18] widely used in diagnostic nuclear medicine, and the cationic $[Tc^{V}O_2$(1,4,8,11-tetraazacyclotetradecane)$]^+$ ($[Tc^{V}O_2(cyclam)]^+$) complex,[17-23] a relatively inert 1:1 complex containing a macrocyclic tetraaza ligand, we decided to test the reactivity of both compounds to 2-mercaptoethanol. We also included the species of pertechnetate in this test. Moreover, we also studied the reactivity of the two complexes of reduced Tc towards reoxidation induced by air or a mild oxidising compound, i.e., copper(II) citrate.[24]

3.2 EXPERIMENTAL

3.2.1 Reagents

All reagents were of analytical grade, and were obtained from Aldrich or Sigma (Bornem, Belgium). Millipore (Milford, MA, USA) Milli-Q Plus water was used for all solution preparations. ^{99}Tc was obtained as ammonium or potassium pertechnetate from Amersham (Amersham, Buckinghamshire, UK). ^{99m}Tc (half-life: 6.02 h; E_γ: 141 keV) was obtained from a Mallinckrodt (Petten, The Netherlands) ^{99}Mo-^{99m}Tc generator.

3.2.2 Chromatography

Single-use Pharmacia (Roosendaal, The Netherlands) PD-10 columns, filled with Sephadex G-25 Medium grade material (particle size range (wet): 90-260 μm; exclusion limit M_r: 5×10^3), were used. Twenty 1-mL fractions were collected. Blue Dextran 2000 (Sigma) was used to determine the interstitial volume V_0 (3.1 mL) of the column. The specifications of the manufacturer listed an internal pore volume V_i of 6.0 mL. All PD-10 columns were eluted with 12 mmol L^{-1} sodium chloride. For some experiments the amount of Tc adsorbed on the column was determined by flushing the column after the experiment with 15 mL 0.3% (w/w) (80 mmol L^{-1}) of hydrogen peroxide.[25]

The elution volume V_e of pertechnetate,[26] the Tc-DTPA complex, and the Tc-cyclam complex under the elution conditions were 13 mL, 4 mL, and 6 mL, respectively. These values correspond with distribution coefficients K_d, which is equal to $(V_e - V_0) / V_i$, of 1.7, 0.2, and 0.5, respectively. The large value of the distribution coefficient K_d of pertechnetate (i.e., 1.7) under the elution conditions corresponds with the gel chromatographic behaviour of anions like I^-, NO_3^-, and ClO_4^-.[27] These anions, like pertechnetate, have a low charge-density, a weak hydration and, consequently, a high lyotropic number.[28] Hydrophobic interaction (a "non-size" effect) between these anions and stationary phase results in distribution coefficients exceeding unity.[29]

3.2.3 Synthesis Tc-DTPA complex

To a 2-mL aliquot of 47.7 MBq L^{-1} (0.77 mmol L^{-1}) $^{99}TcO_4^-$ was added 12 mg (31 μmol) of diethylenetriaminepentaacetic acid (H_5DTPA). Subsequently, 5 mg (19 μmol) of tin(II) tartrate was added. The yellow solution was allowed to stand for 15 min at room temperature and subsequently filtered through a Millex-GV (Millipore) 0.22-μm filter. The formation of the Tc-DTPA complex was tested by loading a 100-μL sample of the filtered solution on a PD-10 column. Apart from the Tc-DTPA complex (98%), only traces of pertechnetate (2%) could be detected. The bulk (1.5 mL) of the filtered Tc-DTPA complex solution was loaded on another PD-10 column. The fraction between 2 and 8 mL (pH 2.8) was collected and subsequently stored at –20°C, until use.

3.2.4 Presence of tin(II)

100 kBq $^{99m}TcO_4^-$ was added to the 2-8 mL fraction in order to get an indication of the presence of residual tin(II) in the Tc-DTPA solution. After 1 h incubation a sample was loaded on a PD-10 column and the amount of remaining $^{99m}TcO_4^-$ was determined.

3.2.5 Synthesis Tc-cyclam complex

To a 2.3-mL aliquot of 87.5 MBq L^{-1} (1.41 mmol L^{-1}) $^{99}TcO_4^-$ was added 10 mg (50 μmol) 1,4,8,11-tetraazacyclotetradecane (cyclam). Subsequently, 15 mg (57 μmol) tin(II) tartrate was added. The pink solution was allowed to stand for 30 min at room temperature and subsequently filtered through a Millex-GV 0.22-μm filter. The formation of the Tc-cyclam complex was tested by loading a 150-μL sample of the filtered solution on a PD-10 column. Apart from the Tc-cyclam complex (99%), only traces of pertechnetate (1%) could be detected. The bulk (1.7 mL) of the filtered Tc-cyclam complex solution was loaded on another PD-10 column. The fraction between 3 and 8 mL (pH 8.0) was collected and subsequently stored at –20°C, until use.

3.2.6 Reactivity Study

3.2.6.1 Reactivity at 20°C. An aliquot of 1.5 mmol (0.12 g) 2-mercapto-ethanol was added to 1 L of 50 mmol L^{-1} HEPES[30] (*N*-[2-hydroxyethyl]-piperazine-*N'*-[2-ethanesulphonic acid]) pH 7.0 (*A*). Aliquots of 2.6 mmol (0.50 g) citric acid and 0.20 mmol (50 mg) $CuSO_4 \cdot 5H_2O$ were added to 1 L of 50 mmol L^{-1} HEPES pH 7.0 (*B*). The reactivity of pertechnetate, the Tc-DTPA and the Tc-cyclam complexes at room temperature (20°C) was studied by diluting their stock solutions 10-fold with *A* or *B* (only Tc-DTPA and Tc-cyclam) in a septum vial and analysing aliquots at set time intervals on a PD-10 column. As control, the Tc-DTPA and the Tc-cyclam stock solutions were diluted with 50 mmol L^{-1} HEPES buffer (pH 7.0) and subjected to an identical analysis procedure.

3.2.6.2 Reactivity at 5°C and –20°C. The reactivity of the Tc-DTPA complex with 2-mercaptoethanol during storage at 5°C or –20°C was studied by diluting the Tc-DTPA solution 10-fold with *A*, followed by a time-dependent analysis. To this end the solution was divided in 25 fractions and stored in screw capped polyethylene tubes. One tube was directly analysed on a PD-10 column. Twelve tubes were kept in a refrigerator at 5°C, and another series of twelve tubes was frozen (–20°C). At set time intervals two tubes were taken from the refrigerator and allowed to warm up for 10 min at room temperature (20°C). Subsequently, aliquots of the tubes were analysed on a PD-10 column. As control, the Tc-DTPA stock solution was diluted with

50 mmol L^{-1} HEPES buffer (pH 7.0) and subjected to an identical analysis procedure.

3.2.7 Detection

^{99}Tc was determined with a Packard (Meriden, CT, USA) Tri-carb 1600 CA, 2750 TR/LL or 4000 Series liquid scintillation analyser, using Ultima Gold XR (Packard) as liquid scintillation cocktail. The efficiency of ^{99}Tc counting under these conditions was 93%. The duration of the counting period was adjusted to bring the uncertainty of the count rate (95% confidence level) below 2%. ^{99m}Tc was determined with a Wallac (Turku, Finland) 1480 Automatic 3" gamma counter, using appropriate energy window settings (104-162 keV).

3.3 RESULTS

3.3.1 Pertechnetate

A 87 μmol L^{-1} solution of TcO_4^- was not reduced by 1.7 mmol L^{-1} 2-mercaptoethanol after 48 h at room temperature in 50 mmol L^{-1} HEPES buffer at pH 7.0. The total amount of Tc recovered was 99 ± 1% (95% confidence limit).

3.3.2 Tc-DTPA complex

The Tc-DTPA complex is inert to oxidation for at least 8 h at room temperature under the experimental conditions (Figure 3.1A). The Tc-DTPA complex remains inert for 1 week when stored at 5°C (Figure 3.1B). Prolonged storage at this temperature leads to the formation of TcO_4^-. No TcO_4^- was formed as a result of oxidation of the complex after one month storage of the Tc-DTPA complex at –20°C (Figure 3.1C). Clearly, the rate of oxidation of the Tc-DTPA complex is dependent on the temperature. All of the $^{99m}TcO_4^-$ added after the filtration and chromatographic purification of the Tc-DTPA synthesis solution was reduced by residual tin(II) after 1 h standing, presumably resulting in the formation of a ^{99m}Tc-DTPA complex. Residual tin(II) could have delayed the oxidation of the Tc-DTPA complex, acting as an anti-oxidant. The total amount of Tc recovered after storage at 20°C, 5°C, and –20°C was 98 ± 1%, 103 ± 2%, and 100 ± 1%, respectively. Storage of the Tc-DTPA complex at

20°C (Figure 3.1D) in the presence of copper(II) citrate resulted in reoxidation and formation of TcO_4^-. Apparently, Tc-DTPA is oxidised by a Fenton type reaction involving copper(I) and (II) ions and molecular oxygen. The total amount of Tc recovered after storage in the presence of copper(II) citrate was 99 ± 1%.

Storage of the Tc-DTPA complex at 20°C (Figure 3.2A), at 5°C (Figure 3.2B), and at –20°C (Figure 3.2C) in the presence of 2-mercaptoethanol resulted in the formation of a new technetium species with a elution volume of 10 mL (K_d = 1.2) under the elution conditions, which was separated from the Tc-DTPA complex. The mechanism of the interaction of 2-mercaptoethanol with the Tc-DTPA complex remains to be elucidated, but a substitution reaction resulting in a $[TcO(SCH_2CH_2O)_2]^-$ complex[15] is a possibility. The total amount of Tc recovered after storage in the presence of 2-mercaptoethanol at 20°C, 5°C, and –20°C was 98 ± 1%, 96 ± 3%, and 98 ± 1%, respectively.

3.3.3 Tc-cyclam complex

The $[TcO_2(cyclam)]^+$ complex remained inert to oxidation by air (Figure 3.3A) or copper(II) citrate (Figure 3.3B) and to added 2-mercapto-ethanol (Figure 3.3C) at room temperature and pH 7.0 for at least 6 h. The total amount of recovered Tc was 98 ± 1%, 97 ± 1%, and 99 ± 1%, respectively.

3.4 DISCUSSION

Our results show that one of the three selected technetium species, i.e., the Tc-DTPA complex, did react with 2-mercaptoethanol. However, the more inert Tc-cyclam complex[17] did not react with 2-mercaptoethanol. The third species, pertechnetate, was not reduced by 2-mercaptoethanol under our experimental conditions (48 h at 20°C and pH 7.0). Apparently, this is due to kinetic reasons, since 2-mercaptoethanol is thermodynamically able to reduce pertechnetate, a weak oxidant, itself.

The data on the technetium speciation in plants is largely based on experiments in which 2-mercaptoethanol, a reducing compound and a ligand for technetium(V) complexes, is applied in the determination procedures. The Tc-DTPA and the Tc-cyclam complexes do not occur in plants. However, the fact

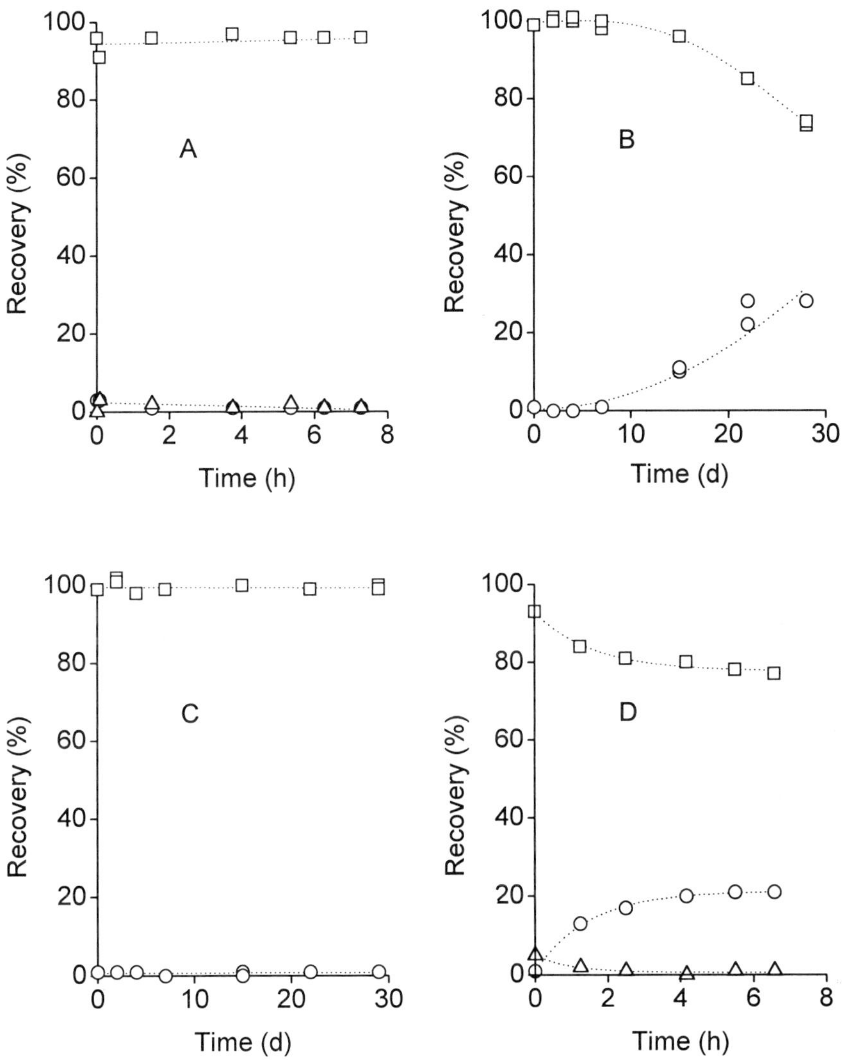

Figure 3.1. Reactivity of the Tc-DTPA complex in 45 mmol L^{-1} HEPES (pH 7.0): (A) at 20°C; (B) at 5°C; (C) at –20°C; (D) at 20°C in the presence of 0.20 mmol L^{-1} Cu(II) citrate. (□) Tc-DTPA, (○) TcO_4^-, and (Δ) column adsorbed Tc.

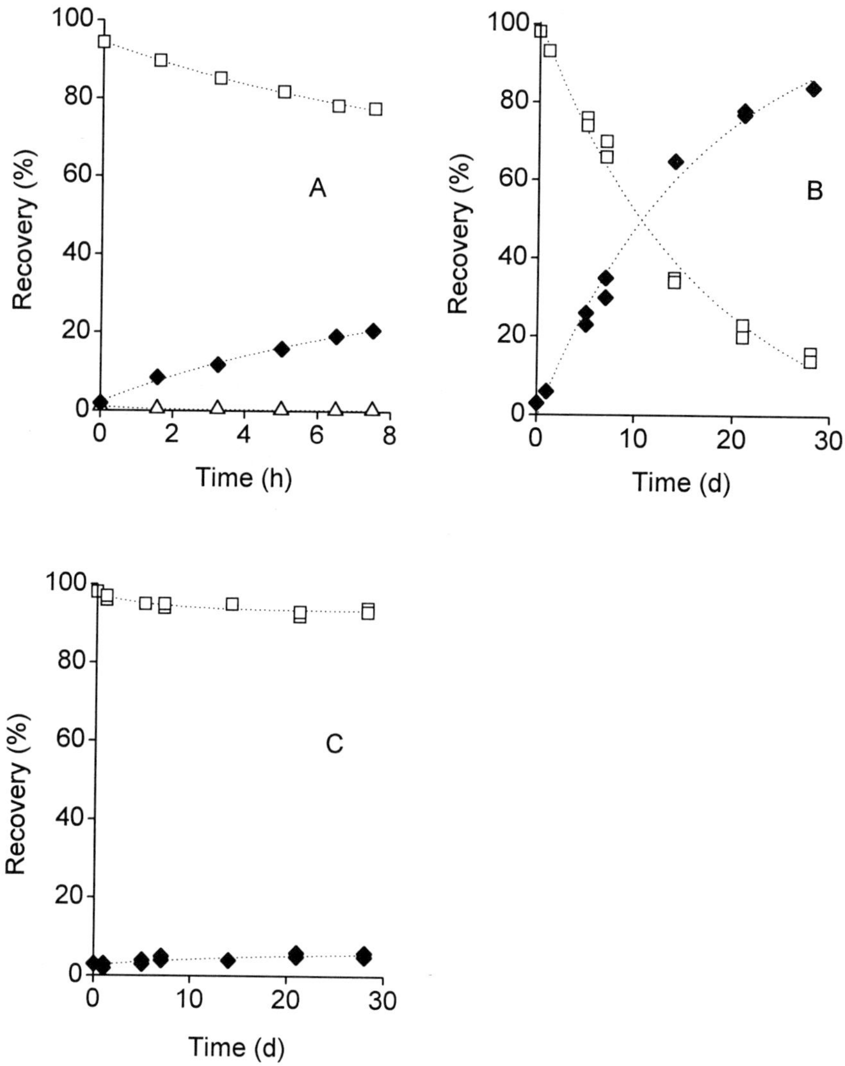

Figure 3.2. Reactivity of the Tc-DTPA complex in the presence of 1.3 mmol L^{-1} 2-mercaptoethanol and 45 mmol L^{-1} HEPES (pH 7.0): (A) at 20°C; (B) at 5°C; (C) at –20°C. (□) Tc-DTPA, (◆) new Tc species, and (Δ) column adsorbed Tc.

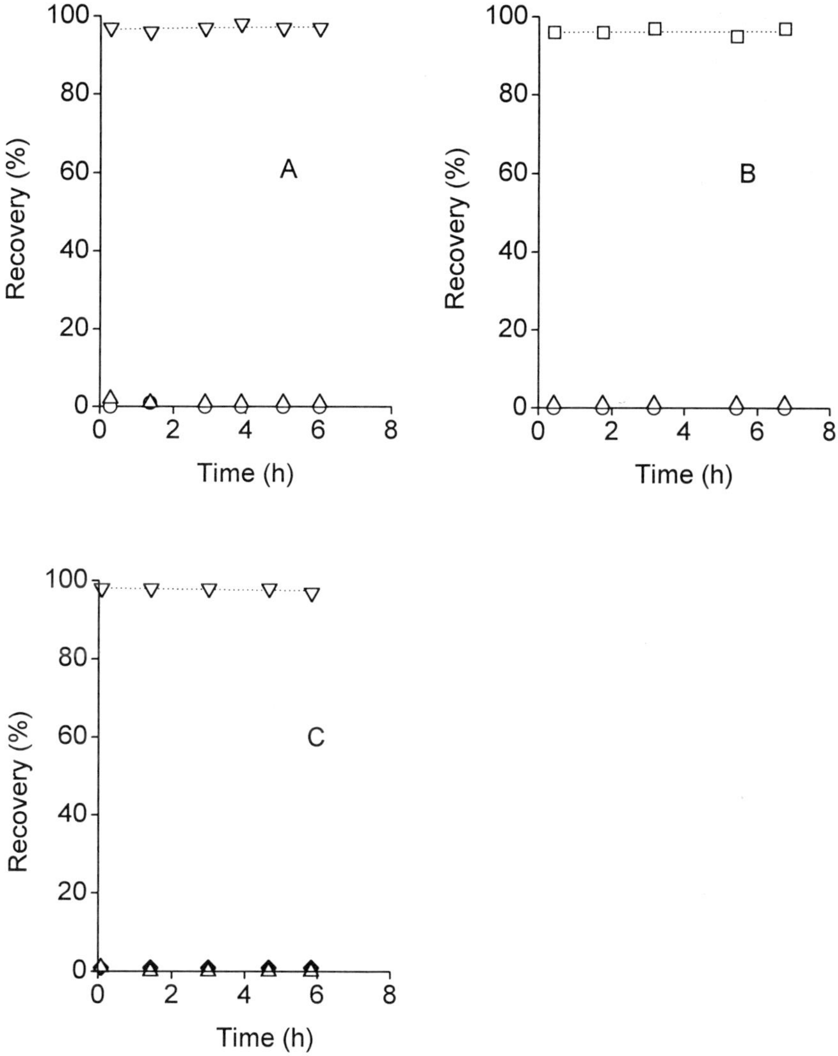

Figure 3.3. Reactivity of the Tc-cyclam complex at 20°C: (A) in 45 mmol L^{-1} HEPES (pH 7.0); (B) in the presence of 0.20 mmol L^{-1} copper(II) citrate in 45 mmol L^{-1} HEPES (pH 7.0); (C) in the presence of 1.6 mmol L^{-1} 2-mercaptoethanol in 45 mmol L^{-1} HEPES (pH 7.0). (∇) Tc-cyclam, (◆) new Tc species, (○) TcO_4^-, and (Δ) column adsorbed Tc.

that a fairly inert complex as the radiopharmaceutical Tc-DTPA did react with 2-mercaptoethanol indicates that labile plant Tc species may also react with 2-mercaptoethanol during the isolation procedure. Therefore, a caveat on the use of 2-mercaptoethanol in plant technetium speciation procedures is justified.

It cannot be ruled out that residual tin(II) or tin(IV) compounds from the complex synthesis might play some role in the reaction of 2-mercaptoethanol with the Tc-DTPA complex. However, tin, among other metal ions, is also present in plant material. Tin(IV) concentrations in plants are generally 0.2 $\mu g\ g^{-1}$ (dry weight),[31] but may go up to 2 mg g^{-1} (dry weight)[32] on contaminated soil. In fact, when dealing with plant technetium speciation studies, not only the effect of 2-mercaptoethanol itself, but also the combined effect of 2-mercaptoethanol and other compounds (e.g., tin and other metal ions) in plant material on technetium species should be considered.

3.5 REFERENCES

(1) Sparkes, S. T.; Long, S. E. *Report AERE-R 12743 DOE/RW/88098*; Harwell Laboratory Oxfordshire, 1988.

(2) Holm, E. *Radiochim. Acta* **1993**, *63*, 57-62.

(3) Wildung, R. E.; McFadden, K. M.; Garland, T. R. *J. Environ. Qual.* **1979**, *8*, 156-161.

(4) Wildung, R. E.; Garland, T. R.; Cataldo, D. (1977) *Health Phys.* **1977**, *32*, 314-317.

(5) Shaw, G.; Bell, J. N. B. in *Plants and the chemical elements*; Farago, M. E., Ed.; VCH: Weinheim, 1994; pp 179-220.

(6) Lembrechts, J. F.; Desmet, G. M.; Overbeek, H. *Environ. Exp. Bot.* **1985**, *25*, 355-360.

(7) Lembrechts, J. F.; Desmet, G. M. *Environ. Exp. Bot.* **1985**, *25*, 361-368.

(8) Lembrechts, J. F.; Desmet, G. M. *Plant Physiol.* **1986**, *81*, 1003-1007.

(9) Lembrechts, J. F.; Desmet, G. M. *Health Phys.* **1989**, *57*, 255-262.

(10) Van Loon, L. R.; Lembrechts, J. F. in *Technetium in the environment*; Desmet, G. M., Myttenaere, C., Eds.; Elsevier: London, 1986, pp 301-306.

(11) Boffey, S. A. in *Methods in Plant Biochemistry, volume 5 Amino acids,*

proteins and nucleic acids; Rogers, L. J., Ed.; Academic press: London, 1991; pp 147-169.
(12) Welander, M. *Physiol. Plant.* **1978**, *43*, 242-246.
(13) Rhodes, B. A. *Nucl. Med. Biol.* **1991**, *18*, 667-676.
(14) Zhang, Z. M.; Ballinger, J. R.; Sheldon, K.; Boxen, I. *Nucl. Med. Biol.* **1992**, *19*, 607-609.
(15) Jones, A. G.; DePamphilis, B. V.; Davison, A. *Inorg. Chem.* **1981**, *20*, 1617-1618.
(16) Kothari, K.; Pillai, M. R. A. *Appl. Radiat. Isot.* **1993**, *44*, 911-915.
(17) Volkert, W. A.; Troutner, D. E.; Holmes, R. A. *Int. J. Appl. Radiat. Isot.* **1982**, *33*, 891-896.
(18) Baldas, J. in *Advances in inorganic chemistry*, Vol. 41; Sykes, A. G., Ed.; Academic press: London, 1994; pp 2-123.
(19) Simon, J.; Troutner, D. E.; Volkert, W. A.; Holmes, R. A. *Radiochem. Radioanal. Lett.*, **1981**, *47*, 111-124.
(20) Zuckman, S. A.; Freeman, G. M.; Troutner, D. E.; Volkert, W. A.; Holmes, R. A.; VanderVeer, D. G.; Kent Barefield, E. *Inorg. Chem.* **1981**, *20*, 2386-2389.
(21) Troutner, D. E.; Simon, J.; Ketring, A. R.; Volkert, W.; Holmes, R. A. *J. Nucl. Med.* **1980**, *21*, 443-448.
(22) Riché, F.; Vidal, M.; Bolzati, C.; Uccelli, L.; Marchi, A.; Duatti, A. *Inorg. Chim. Acta* **1995**, *231*, 147-151.
(23) Simon, J. *Ph.D. thesis*; University of Missouri: Colombia, 1981.
(24) Berger, R. *Int. J. Appl. Radiat. Isot.* **1982**, *33*, 1341-1344.
(25) Muddukrishna, S. N.; Jorge, P.; Machner, W.; Sykes, T. R.; Noujaim, A. A. *Appl. Radiat. Isot.* **1994**, *45*, 1009-1019.
(26) Persson, B. R. R.; Darte, L. *J. Chromatogr.* **1974**, *101*, 315-326.
(27) Borák, J. *J. Chromatogr.* **1978**, *155*, 69-82.
(28) Deguchi, T.; Hisanaga, A.; Nagai, H. *J. Chromatogr.* **1977**, *133*, 173-179.
(29) Verbeke, K.; Verbruggen, A. *J. Pharm. Biomed. Anal.* **1996**, *14*, 1209-1213.
(30) Perrin, D. D.; Dempsey, B. *Buffers for pH and Metal Ion Control*; Chapman and Hall: London, 1974; pp 30 and 60.
(31) Farago, M. E.; Mehra, A. in *Plants and the chemical elements*; Farago,

M. E., Ed.; VCH: Weinheim, 1994; pp 241-264.
(32) Kabata-Pendias, A.; Pendias, H. in *Trace elements in soils and plants*; CRC Press: Boca Raton, 1985; p 154.

Chapter 4

Technetium Speciation: Non-size Effects in Size-Exclusion Chromatography*

4.1 INTRODUCTION

Technetium, with atomic number 43, is the lightest element that has only radioactive isotopes. From an environmental point of view, the most important isotope is ^{99}Tc, a β^--emitter with $E_{\beta^-,max}$ = 292 keV and a half-life of 2.13×10^5 years; it is a fission product of ^{235}U and ^{239}Pu. The main sources of ^{99}Tc to the environment are the nuclear power generation and associated industry, nuclear weapon testing programs, and to a lesser degree, diagnostic nuclear medicine.[1,2] The high solubility and the mobility in soil of the pertechnetate anion (TcO_4^-), the dominant chemical form of technetium in aerobic environments,[3,4] combined with the large soil-to-plant transfer factor of technetium,[3,5] have stimulated research in the radioecology of this element. The available data[6-9] suggest the uptake and metabolism of pertechnetate in plants to, so far unidentified, compounds containing reduced technetium. Aqueous size-exclusion chromatography, with its mild separation conditions, is predominantly used to separate these, potentially labile, compounds.[6-11]

In ideal size-exclusion chromatography, a separation results from the distribution of a solute between the moving mobile phase and the stagnant

* This chapter is a slightly modified version of the paper "Technetium Speciation: Non-size Effects in Size-Exclusion Chromatography"; Harms, A. V.; Van Elteren, J. T.; Claessens, H. A. *J. Chromatogr. A* **1996**, *755*, 219-225.

portion of the mobile phase, retained within the porous structure of the stationary phase. However, apart from these sieving effects, interaction of technetium compounds with the stationary phase (non-size effects) was shown by studies concerning the environmental behaviour of ^{99}Tc or the development of ^{99m}Tc compounds used as imaging agents in diagnostic nuclear medicine.[12-25]

For pertechnetate large elution volumes and, consequently, high distribution coefficients (up to 29) were observed on Sephadex material.[12] Interaction of labile technetium complexes with size-exclusion material results in degradation or column adsorption of these complexes.[13-25] For instance, technetium compounds, isolated from spinach plants grown on nutrient solutions spiked with $NH_4{}^{99}TcO_4$, were adsorbed to a substantial degree (up to 80%) on several types of columns.[6]

Generally, non-size effects in size-exclusion chromatography comprise adsorption, ion-exclusion,[26] hydrophobic interaction,[27] and ion-exchange.[28] To evaluate the influence of these non-size effects on technetium speciation, we investigated the behaviour of three technetium compounds: the pertechnetate anion (TcO_4^-), the anionic Tc-diethylenetriaminepentaacetate complex (Tc-DTPA)[29-31] and the cationic $[Tc^VO_2$(1,4,8,11-tetraazacyclotetradecane)$]^+$ complex (Tc-cyclam)[29] on three different types of size-exclusion packing material: a soft polydextran gel (Sephadex G-25), a silica-based hydrophilic packing (Zorbax GF-250), and a rigid polymer gel (HEMA-SEC BIO 1000). The influence of the composition of the mobile phase and the column temperature on the selectivity of these chromatographic systems, caused by the non-size effects, has been studied.

4.2 EXPERIMENTAL

4.2.1 Reagents

All reagents used are described in section 3.2.1.

4.2.2 Synthesis Tc-DTPA

The synthesis of the Tc-DTPA complex is described in section 3.2.4.

4.2.3 Synthesis Tc-cyclam

The synthesis of the Tc-cyclam complex is described in section 3.2.6.

4.2.4 HPLC-apparatus

The HPLC-apparatus used included: a Bio-Rad (Hercules, CA, USA) Series 5000T System equipped with a Model 2700 solvent delivery system containing titanium pump heads, a Rheodyne 7125T titanium injector (sample loop 20 μL), a Bio-Rad HRLC System Interface, an IBM compatible 286 PC with Microsoft (Redmond, WA, USA) Windows 3.1 based Series 800 HRLC System v. 2.30 software, and a Model 2100 fraction collector. A Shimadzu (Kyoto, Japan) Model SPD-10AV variable wavelength (190-900 nm) UV-Vis detector, with an inert PCTFE flow cell, was used to measure the absorbance at 245 nm. To control the temperature of the column, an Eppendorf CH-430 column heater and an Eppendorf TC-50 temperature controller (Alltech, Deerfield, IL, USA) were used. All tubing, frits, and fittings were made of titanium or PEEK (polyether ether ketone). The mobile phase was degassed with helium, and filtered through a 0.22-μm Millex-GV filter (Millipore), prior to use. The extinction coefficient ε of pertechnetate at 245 nm, was determined to be 6.2×10^3 L mol^{-1} cm^{-1}.

4.2.5 HEMA Column

An Alltech MF-Plus (Metal-Free) HEMA-SEC BIO 1000 size-exclusion column was used (250 mm × 4.6 mm i.d.; total volume V_c: 4.15 mL; particle size: 10 μm; exclusion limit M_r: 1.2×10^6 - 1.8×10^6). The packing material is a macroporous hydrophilic copolymer of 2-hydroxyethylmethacrylate and ethylene dimethacrylate. Blue Dextran 2000 was used to determine the interstitial volume V_0 (1.6 mL) of the column. The specifications of the manufacturer listed an internal pore volume V_i of 1.1 mL. A MF-Plus Guard column (Alltech) was used as pre-column (10 mm × 4.6 mm i.d.). The flow rate used was 1 mL min^{-1}.

4.2.6 Zorbax Column

A Rockland (Newport, DE, USA) Zorbax Bio Series GF-250 size-exclusion column was used (250 mm × 9.4 mm i.d.; total volume V_c: 17.3 mL; particle size: 5 μm; pore size: 150 Å; exclusion limit M_r: 0.4×10^6). The zirconium stabilised packing material has been made hydrophilic by covalently binding an organosilane reagent to the silica support. Blue Dextran 2000 was used to determine the interstitial volume V_0 (5.9 mL) of the column. The specifications of the manufacturer listed an internal pore volume V_i of 6.3 mL. A Zorbax Diol guard column (Rockland) was used as pre-column (12.5 mm × 4 mm i.d.).

4.2.7 PD-10 (Sephadex) Column

The single-use PD-10 columns used are described in section 3.2.2.

4.2.8 Detection of ^{99}Tc

The detection of ^{99}Tc proceeded as described in section 3.2.7.

4.2.9 Methods

Retention characteristics in size-exclusion chromatography can be represented by the distribution coefficient $K_d = (V_e - V_0) / V_i$, where V_e, V_0, and V_i refer to the elution volume of the solute, the column interstitial volume (mainly the liquid volume between the packing particles), and the internal pore volume, respectively. This distribution coefficient K_d can be represented as the product of K_{SEC}, the size-exclusion distribution coefficient, and K_{LC}, the distribution coefficient arising from solute-surface interactions. K_{SEC} is constrained to values between 0 and 1, representing the extremes of complete exclusion and permeability of the pore volume by the solute.

4.3 RESULTS

4.3.1 HEMA

For the polymer-based HEMA material the distribution coefficient K_d of pertechnetate is strongly dependent on the sodium chloride concentration of the mobile phase and the column temperature (Figure 4.1A). The use of pure water as a mobile phase results in exclusion of pertechnetate from the column pores and, consequently, in a negligible distribution coefficient. Increasing the sodium chloride concentration of the mobile phase results in significantly higher distribution coefficients. It can be seen that elevation of the temperature

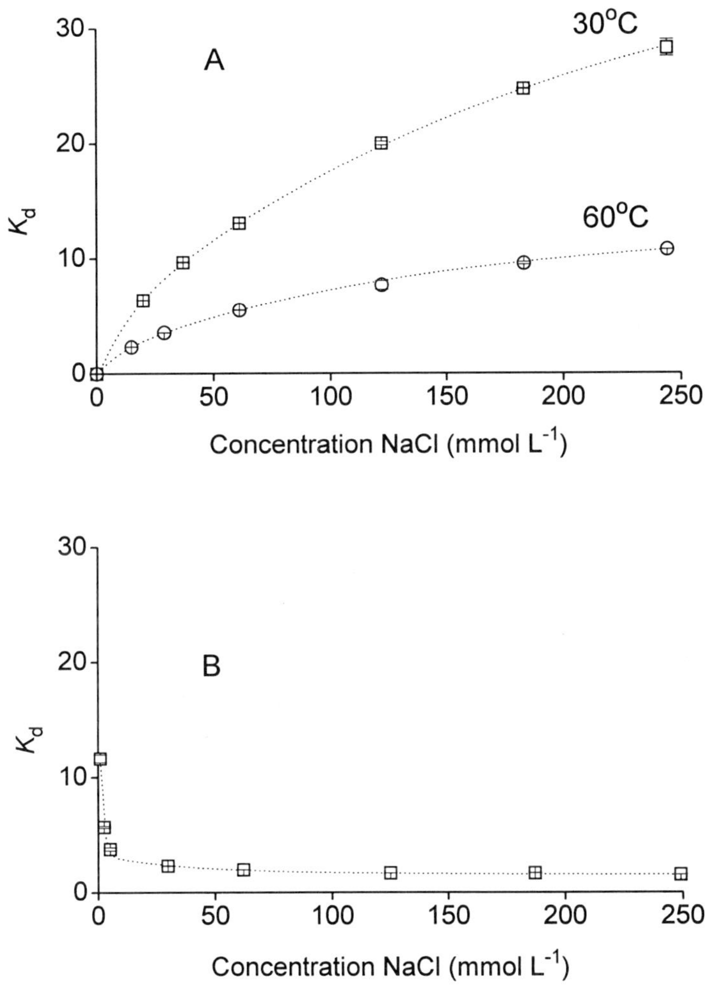

Figure 4.1. The distribution coefficient K_d (with standard error) of TcO_4^- as a function of the NaCl concentration of the mobile phase and the column temperature. Mobile phase: water containing several concentrations NaCl; detection: 245 nm absorbance and ^{99}Tc activity; flow rate: 1 mL min^{-1}. (A) Column: HEMA; column temperature: 30°C (□) or 60°C (○). (B) Column: Zorbax; column temperature: 30°C.

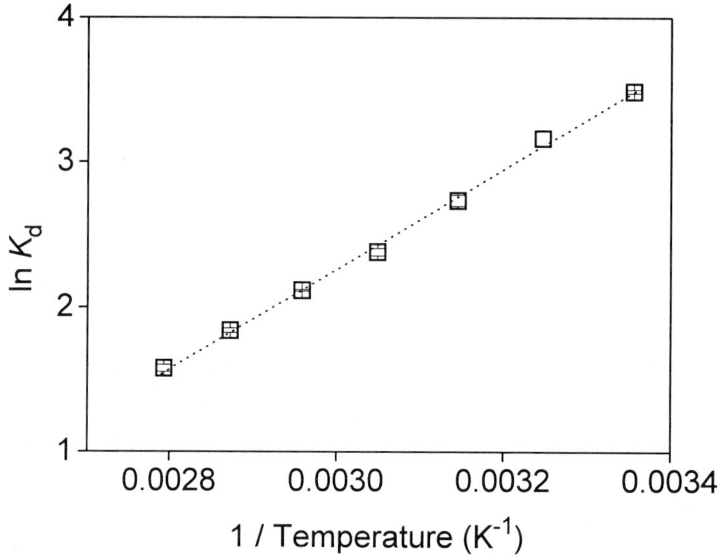

Figure 4.2. Van 't Hoff plot (the logarithm of the distribution coefficient K_d (with standard error) of TcO_4^- as a function of the inverse of the column temperature). Column: HEMA; mobile phase: water containing 250 mmol L^{-1} NaCl; detection: 245 nm absorbance and ^{99}Tc activity; flow rate: 1 mL min^{-1}. Slope ± S.D.: $3.44 \times 10^3 \pm 30$; intercept ± S.D.: –8.1 ± 0.1; correlation coefficient: 0.9988; n = 14.

results in lower distribution coefficients. The temperature dependence of the distribution coefficient of pertechnetate was studied between 25 and 85°C, in steps of 10°C, using a mobile phase of 250 mmol L^{-1} NaCl. A van 't Hoff plot was constructed (Figure 4.2), showing a linear relationship between the logarithm of the distribution coefficient and the reciprocal of the column temperature. The slope of the line, $(-\Delta H° / R)$, gives for the standard enthalpy of the retention process $\Delta H°$ a value of –29 kJ mol^{-1}, showing the exothermic nature of the interaction between pertechnetate and the stationary phase.

The effect of the sodium chloride concentration of the mobile phase on the distribution coefficient at 60°C of pertechnetate, Tc-DTPA, and Tc-cyclam is shown in Figure 4.3. As can be seen from this figure, no separation was obtained between pertechnetate and Tc-cyclam at a NaCl concentration of 8 mmol L^{-1}. For optimum selectivity, the NaCl concentration should be 5 mmol L^{-1} or exceeding 12 mmol L^{-1}. Separation at these NaCl concentrations is feasible, as illustrated in Figure 4.4. Quantitative recovery (>98%) was observed for all technetium compounds.

4.3.2 Sephadex

On Sephadex columns similar trends in the distribution coefficient dependence as for the HEMA column are observed (Table 4.1), although to a lesser extent. The distribution coefficients of pertechnetate and Tc-cyclam are dependent on the sodium chloride concentration of the mobile phase, while for Tc-DTPA only a small influence of the composition of the mobile phase on the distribution coefficient is observed.

Table 4.1. The 95% confidence limits of the distribution coefficient K_d of pertechnetate, Tc-DTPA, and Tc-cyclam as a function of the sodium chloride concentration of the mobile phase on Sephadex G-25 column at 20°C. Detection: ^{99}Tc activity; injection: 100 μL.

	TcO_4^-	Tc-cyclam	Tc-DTPA
water	0.6 ± 0.1	1.7 ± 0.03	0.2 ± 0.1
5 mmol L^{-1} NaCl	1.4 ± 0.2	0.8 ± 0.1	0.2 ± 0.1
12 mmol L^{-1} NaCl	1.6 ± 0.1	0.5 ± 0.04	0.3 ± 0.1

4.3.3 Zorbax

The use of water as a mobile phase results in the adsorption of pertechnetate on the silica-based Zorbax column. Addition of sodium chloride to the mobile phase results in a decrease of the distribution coefficient (Figure 4.1B). Other compositions of the mobile phase result in comparable distribution coefficients for pertechnetate: K_d = 2.1 for 50 mmol L^{-1} HEPES buffer (pH 7.0; 30°C),

K_d = 1.5 for 50 mmol L^{-1} Na-HEPES buffer and 0.88 mol L^{-1} NaCl (pH 7.0; 60°C), and K_d = 3.0 for 0.12 mol L^{-1} NaOAc buffer (pH 4.4; 30°C).

Tc-DTPA and Tc-cyclam are partly and almost completely adsorbed, respectively, on the column using 50 mmol L^{-1} Na-HEPES buffer (pH 7.0; 30°C) as mobile phase. The distribution coefficient for the Tc-DTPA complex is 1.3. The recovery of Tc-DTPA is between 68 and 90%. An injection of 0.3% (w/w) H_2O_2, to oxidise adsorbed Tc, results in an additional recovery of 10 to 15% technetium. Neither the addition of sodium chloride [using 25 mmol L^{-1} Na-HEPES (pH 7.0) buffer and 0.24 mol L^{-1} NaCl as mobile phase], nor the addition of methanol [using 45 mmol L^{-1} Na-HEPES (pH 7.0) and 10% (v/v) MeOH as mobile phase], results in the elution of Tc-cyclam.

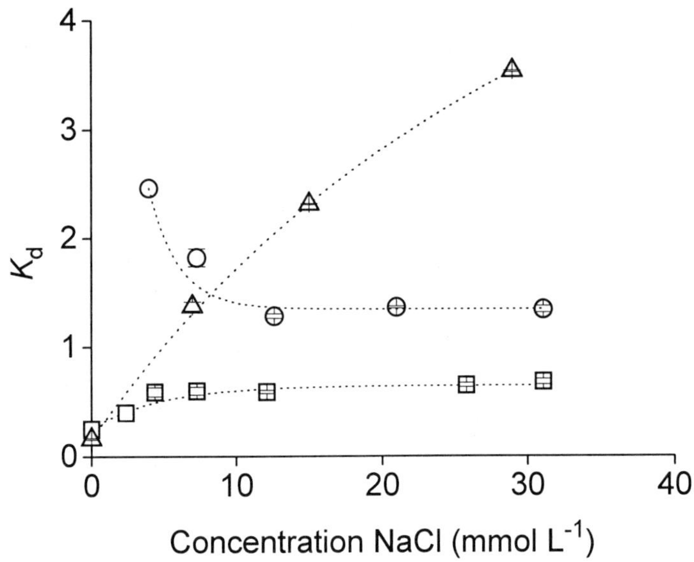

Figure 4.3. The distribution coefficients K_d (with standard errors) of TcO_4^-, Tc-DTPA, and Tc-cyclam as a function of the NaCl concentration of the mobile phase. Column: HEMA; column temperature: 60°C; detection: 245 nm absorbance and ^{99}Tc activity; flow rate: 1 mL min^{-1}. (Δ) TcO_4^-, (□) Tc-DTPA, and (O) Tc-cyclam.

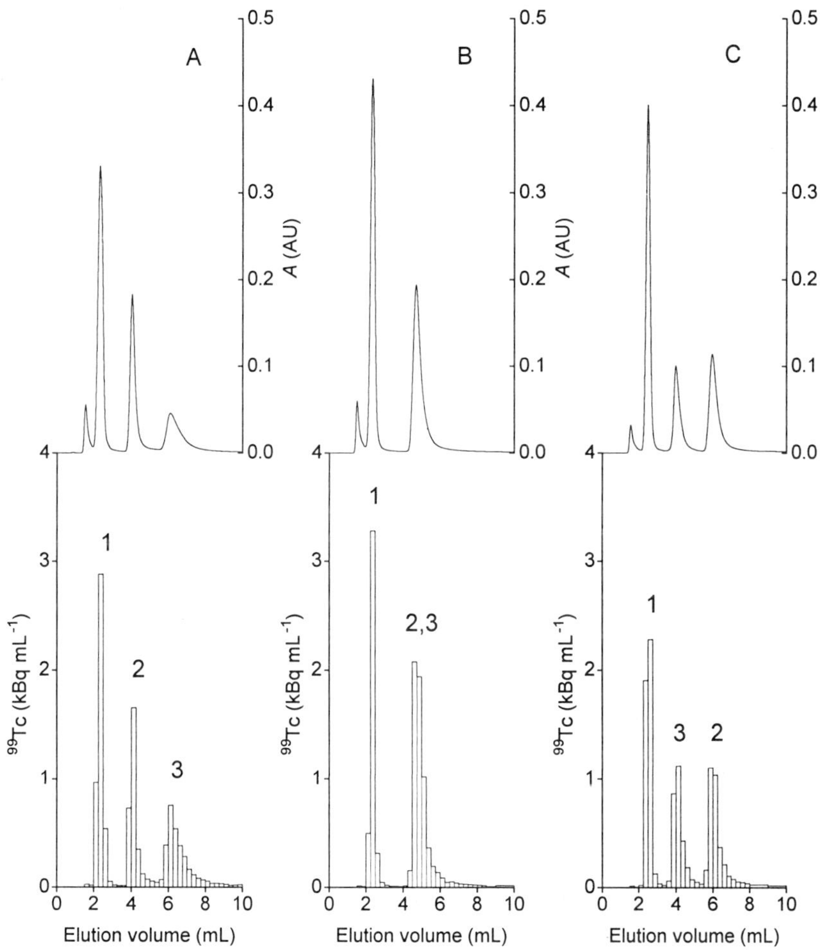

Figure 4.4. Separation of Tc-DTPA (1), TcO_4^- (2) and Tc-cyclam (3). Column: HEMA; column temperature: 60°C; detection: 245 nm absorbance (*top*) and ^{99}Tc activity (*bottom*); flow rate: 1 mL min^{-1}. Mobile phase: (A) water containing 5 mmol L^{-1} NaCl, (B) water containing 8 mmol L^{-1} NaCl, and (C) water containing 13 mmol L^{-1} NaCl.

4.4 DISCUSSION AND CONCLUSION

As has been pointed out for other charged solutes,[33] the observed chromatographic behaviour of the pertechnetate anion, the anionic Tc-DTPA complex, and the cationic Tc-cyclam complex, cannot be fully explained in terms of the sieving effect based on their ionic sizes. Interactions between the technetium compounds and the stationary phase, i.e., non-size effects, must also be taken into consideration, although their significance is greatly dependent on the properties of the applied solutes and stationary and mobile phases.[34]

Ion exclusion,[12] a non-size effect, results in the faster elution of an anionic solute than can be expected from its ionic size. For anions this effect might be observed when a low ionic strength mobile phase is used. This effect is the result of the negative charge of the column material at neutral pH, caused by the presence of a small number of carboxyl or silanol groups. This is demonstrated by the exclusion of pertechnetate on Sephadex and HEMA material, using water as mobile phase, and the distribution coefficient of the anionic Tc-DTPA complex for Sephadex and HEMA material, which turns out to be smaller (i.e., 0.2-0.5) than expected for a solute of its size. The distribution coefficient of Tc-DTPA is only slightly affected by the addition of sodium chloride to the mobile phase. However, the distribution coefficient of Tc-DTPA on silica-based Zorbax material was larger (i.e., 1.3) than expected for a solute of its size.

Hydrophobic interaction,[18] another non-size effect that causes more retention than would be expected from the size of a solute, is observed for pertechnetate on Sephadex and HEMA material, when salt has been added to the mobile phase. Borák[35] suggested that hydroxyethyl methacrylate material can be used for the separation of anions. This author claims that differences in the chromatographic behaviour of anions are caused by the degree and type of water arrangement in the vicinity of the anions and the surface. Low charge-density structure-breaking anions, e.g., I^- and SCN^-, are adsorbed on the gel surface, whereas structure-forming anions, e.g., SO_4^{2-} and F^-, are squeezed out of the pores of the gel. The elution sequence of anions corresponds to the lyotropic number (which is derived from the concentration of sodium salts of different anions required to flocculate agar)[36] of the anions, while the adsorption of ions is increased if the mobile phase contains an electrolyte.

Pertechnetate, like anions such as NO_3^-, I^-, SCN^-, ReO_4^-, and ClO_4^-, has a low charge-density, a weak hydration, and, consequently, a high lyotropic number. The dependence of the distribution coefficient of pertechnetate as a function of the ionic strength of the mobile phase, and the increase of retention with decreasing temperatures on the HEMA material, corresponds to the gel chromatographic behaviour of other anions with a low charge-density.[37,38]

A third non-size effect, cation exchange,[21] is observed for Tc-cyclam on the Sephadex and HEMA columns when water was used as a mobile phase. Increasing the sodium chloride concentration of the mobile phase, results in a decrease of the distribution coefficients.

The interaction of pertechnetate with Zorbax material, silica particles with a zirconium oxide cladding, covalently bound to a hydrophilic organo-silane,[39] is contradictory to its behaviour on the other packings. When pure water or low concentration sodium chloride solutions were used as mobile phase, adsorption or high distribution coefficients, respectively, were observed. An increase in the sodium chloride concentration of the mobile phase resulted in a decreasing distribution coefficient (down to 1.5). An explanation for this behaviour might be found in the influence of the zirconium oxide cladding. Hydrous zirconium oxide, like alumina, shows, depending on the pH, both anion (low pH) and cation (high pH) exchange behaviour.[40]

Finally, a fourth non-size effect, adsorption, is observed for Tc-cyclam on the silica-based Zorbax column. This result is in accordance with the unsuccessful attempts by Hoffman et al.[41] to develop a separation method for technetium tetraamine complexes on octadecylsilica-bonded column packings. They attributed this behaviour of the complexes to the negatively charged surface of the silica at $pH > 2$ that strongly interacts with these complexes.

Thus, size-exclusion chromatography provides the chromatographer with several variables (e.g., ionic strength and temperature) to control the retention behaviour of, especially, pertechnetate and the Tc-cyclam complex, thus optimising the separation between technetium compounds. Several non-size effects (i.e., adsorption, ion-exclusion, hydrophobic interaction, and cation exchange) play a dominant role in the separation of pertechnetate, the Tc-DTPA complex, and the Tc-cyclam complex, using a size-exclusion packing. The HEMA column shows the best separation characteristics for this type of separation, because of the absence of column-adsorption and, consequently, complete recovery of technetium compounds.

4.5 REFERENCES

(1) Sparkes, S. T.; Long, S. E. *Report AERE-R 12743 DOE/RW/88098*; Harwell Laboratory: Oxfordshire, 1988.
(2) Holm, E. *Radiochim. Acta* **1993**, *63*, 57-62.
(3) Wildung, R. E.; Garland, T. R.; Cataldo, D. A. *Health Phys.* **1977**, *32*, 314-317.
(4) Wildung, R. E.; McFadden, K. M.; Garland, T. R. *J. Environ. Qual.* **1979**, *8*, 156-161.
(5) Shaw, G.; Bell, J. N. B. in *Plants and the chemical elements*; Farago, M. E., Ed.; VCH: Weinheim, 1994; p 179-220.
(6) Lembrechts, J. F.; Desmet, G. M.; Overbeek, H. *Environ. Exp. Bot.* **1985**, *25*, 355-360.
(7) Van Loon, L. R.; Desmet, G. M.; Cremers, A. *Health Phys.* **1989**, *57*, 309-314.
(8) Lembrechts, J. F.; Desmet, G. *Health Phys.* **1989**, *57*, 255-262.
(9) Cataldo, D. A.; Garland, T. R.; Wildung, R. E. in *Technetium in the environment*; Desmet, G. M., Myttenaere, C., Eds.; Elsevier: London, 1986; pp 265-280.
(10) Garten, Jr., C. T.; Lomax, R. D. *Health Phys.* **1989**, *57*, 299-307.
(11) Henrot, J. *Health Phys.* **1989**, *57*, 239-245.
(12) Persson, B. R. R.; Darte, L. *J. Chromatogr.* **1974**, *101*, 315-326.
(13) Eckelman, W.; Meinken, G.; Richards, P. *J. Nucl. Med.* **1971**, *12*, 596-600.
(14) Eckelman, W.; Richards, P. *J. Nucl. Med.* **1972**, *13*, 202-204.
(15) Valk, P. E.; Dilts, C. A.; McRae, J. *J. Nucl. Med.* **1973**, *14*, 235-237.
(16) Steigman, J.; Williams, H. P.; Richards, P.; Lebowitz, E.; Meinken, G. *J. Nucl. Med.* **1974**, *15*, 318-319.
(17) Valk, P. E.; McRae, J. *J. Nucl. Med.* **1974**, *15*, 319.
(18) Eckelman, W. C.; Levenson, S. M. *Int. J. Appl. Radiat. Isot.* **1977**, *28*, 67-82.

(19) Kroesbergen, J.; Wortelboer, M. R.; Rijnsent, A.; Gelsema, W. J.; De Ligny, C. L. *Nucl. Med. Biol.* **1987**, *14*, 37-41.
(20) Liang, F. H.; Virzi, F.; Hnatowich, D. J. *Nucl. Med. Biol.* **1987**, *14*, 63-67.
(21) Liang, F. H.; Virzi, F.; Hnatowich, D. J. *Nucl. Med. Biol.* **1987**, *14*, 555-561.
(22) Zhou, T.; Hirth, W. W.; Heineman, W. R.; Deutsch, E. *Nucl. Med. Biol.* **1988**, *15*, 493-498.
(23) Rhodes, B. A. *Nucl. Med. Biol.* **1991**, *18*, 667-676.
(24) Zamora, P. O.; Mercer-Smith, J. A.; Marek, M. J.; Schulte, L. D.; Rhodes, B. A. *Nucl. Med. Biol.* **1992**, *19*, 797-802.
(25) Muddukrishna, S. N.; Jorge, P.; Machner, W.; Sykes, T. R.; Noujaim, A. A., *Appl. Radiat. Isot.* **1994**, *45*, 1009-1019.
(26) Shibukawa, M.; Ohta, N. in *Aqueous size-exclusion chromatography*; Dubin, P. L., Ed.; Elsevier: Amsterdam, 1988; p 77-115.
(27) Janado, M. in *Aqueous size-exclusion chromatography*; Dubin, P. L., Ed.; Elsevier: Amsterdam, 1988; p 23-54.
(28) Dubin, P. L. in *Aqueous size-exclusion chromatography*; Dubin, P. L., Ed.; Elsevier: Amsterdam, 1988; p 55-75.
(29) Volkert, W. A.; Troutner, D. E.; Holmes, R. A. *Int. J. Appl. Radiat. Isot.* **1982**, *33*, 891-896.
(30) Kothari, K.; Pillai, M. R. A. *Appl. Radiat. Isot.* **1993**, *44*, 911-915.
(31) Baldas, J. in *Advances in inorganic chemistry*, Vol. 41; Sykes, A. G. Ed.; Academic press: London, 1994; p 2-123.
(32) Harms, A. V.; Wolterbeek, H. Th.; Woroniecka, U. D. *Appl. Radiat. Isot.* **1996**, *47*, 605-610.
(33) Shibukawa, M.; Ohta, N.; Kuroda, R. *Anal. Chem.* **1981**, *53*, 1620-1627.
(34) Verbeke, K.; Verbruggen, A. *J. Pharm. Biomed. Anal.* **1996**, *14*, 1209-1213.
(35) Borák, J. *J. Chromatogr.* **1978**, *155*, 69-82.
(36) McBain, J. W. *Colloid Science*; Reinhold: New York, 1950; p 131.
(37) Ortner, H. M. *J. Chromatogr.* **1975**, *107*, 335-339.
(38) Deguchi, T.; Hisanaga, A.; Nagai, H. *J. Chromatogr.* **1977**, *133*, 173-179.
(39) Stout, R. W.; DeStefano, J. J. *J. Chromatogr.* **1985**, *326*, 63-78.

(40) Amphlett, C. B. *Inorganic ion exchangers*; Elsevier: Amsterdam, 1964.
(41) Hoffman, T. J.; Volkert, W. A.; Troutner, D. E.; Holmes, R. A. *Int. J. Appl. Radiat. Isot.* **1984**, *35*, 223-225.

Chapter 5

A Dual Radiotracer Technique for Technetium Speciation with Emphasis on Probing of Artefacts: a Case Study for Spinach Plants*

5.1 INTRODUCTION

In recent years the focus in trace element studies in biological and environmental sciences has shifted from total element determination to element speciation. Element speciation is defined as the identification and quantification of different chemical forms of an element, i.e., oxidation state and type and number of substituents. It is required for obtaining information on chemical reactivity, toxicity, and bioavailability in biological and environmental systems. In general, interfacing of a high-resolution separation technique with an element-specific quantification technique is required for optimal information retrieval. Separation techniques include gas[1,2] and liquid[3] chromatography, while element-specific quantification techniques comprise several modes of atomic spectrometry[4] and mass spectrometry.[5,6]

---

* This chapter is a slightly modified version of the paper "A Dual Radiotracer Technique for Technetium Speciation with Emphasis on Probing of Artifacts: a Case Study for Spinach Plants"; Harms, A. V.; Van Elteren, J. T.; Wolterbeek, H. Th.; De Goeij, J. J. M. **1998** In preparation.

Although the technical aspects of speciation procedures are relatively well-developed using hyphenated techniques, in general, it is difficult to predict what will happen to the integrity of the species due to storage, pretreatment, and the analytical method itself. Moreover, it is not always easy to follow the uptake, transformation, and emission of species in biological and environmental systems. Radiotracers may be used to trace the species and to monitor species alteration during the analytical pretreatment procedure;[7] with a dual radiotracer both requirements can be fulfilled in one single experiment. The basic assumption is that both radiotracers should have identical properties as the species under study, i.e., the tracee.

In this work a novel dual radiotracer technique both for tracing of ^{99}Tc and probing speciation artefacts is presented and illustrated with a study on speciation of ^{99}Tc in spinach plants. Technetium is the lightest element that has only radioactive isotopes. From an environmental point of view, the isotope ^{99}Tc is the most important one; it is one of the fission products of ^{235}U and ^{239}Pu. The main sources of ^{99}Tc release to the environment are nuclear power generation and associated industry, nuclear weapon testing programs, and to a much lesser degree, diagnostic nuclear medicine.[8] ^{99}Tc, being an artificial element, in the form of pertechnetate ($Tc^{VII}O_4^-$), is of radioecological interest, because of its extraordinary high mobility in soil and its ability to concentrate in plants.[9-11] Due to the fact that ^{99}Tc emits only β^--particles, it is difficult to determine; normally, laborious radiochemical separation methods are necessary prior to liquid scintillation counting. To allow easy tracing of ^{99}Tc during cultivation of spinach plants, the γ-emitting radiotracers $^{95m}TcO_4^-$ or $^{99m}TcO_4^-$ were added to the substrate containing $^{99}TcO_4^-$ (see Table 2.1 for nuclear data of technetium isotopes). The results of studies performed so far on speciation of technetium in plants suggest a metabolic transformation of pertechnetate in plants to, so far unidentified, species containing reduced technetium (TcX) with relatively low molecular masses.[12-15] However, because of the absence of sufficient checks to relate the results of the analytical procedures used to the actual species distribution in the plants, these results should be interpreted with caution.[16] To this end, a possible species alteration was controlled by addition of either $^{99m}TcO_4^-$ or ^{95m}TcX to monitor procedural losses and redox conversions during the analytical pretreatment procedure. A mild size exclusion chromatography separation procedure as developed before[17] was used for separation of TcO_4^- and TcX. A mathematical treatment is given to derive the correction factors involved.

5.2 EXPERIMENTAL

5.2.1 Reagents

All reagents used are described in section 3.2.1. ^{95m}Tc was obtained from Los Alamos National Laboratory (Los Alamos, NM, USA). The ^{95m}Tc stock solution also contained ^{97m}Tc (half-life 91 d), which decays to ^{97}Tc (half-life 2.6 $\times 10^6$ y). The presence of these two isotopes presented no detection problems of ^{95m}Tc or ^{99m}Tc.

5.2.2 HPLC-apparatus

The HPLC-apparatus used is described in section 4.2.4.

5.2.3 Column and Separation Conditions

The HEMA-SEC BIO 1000 size-exclusion column is described in section 4.2.5. A buffer (70 mmol L^{-1} HEPES and 8 mmol L^{-1} NaCl; pH 7.0) was used as mobile phase. The column temperature used was 60°C.

5.2.4 Cultivation of Plants

Seeds of spinach plants, *Spinacea oleracea L.*, were germinated on moist filtration paper for 10 d in the dark. Subsequently, 30 seedlings were grown on 30 L aerated modified Hoagland[18] nutrient solution, which contained 0.6 mmol L^{-1} $Ca(NO_3)_2$, 0.8 mmol L^{-1} KNO_3, 0.19 mmol L^{-1} KH_2PO_4, 0.26 mmol L^{-1} $MgSO_4$, 2 μmol L^{-1} $MnSO_4$, 0.85 μmol L^{-1} $ZnSO_4$, 0.15 μmol L^{-1} $CuSO_4$, 0.02 mmol L^{-1} H_3BO_3, 0.25 μmol L^{-1} Na_2MoO_4, and 0.041 mmol L^{-1} NaFe(EDTA), at pH 5.8. The plants were kept in a climate room at 70% relative humidity with a day-night regime of 16-8 h at 25°C and 20°C, respectively. The light intensity was 100 μE m^{-2} s^{-1}.

5.2.5 Incubation of Plants with TcO_4^-

After 10 d of growth, the plants were weighed and placed in 100 mL fresh nutrient solution (see section 5.2.4) containing 0.85 μmol L^{-1} $^{99}TcO_4^-$, spiked with either 250 Bq mL^{-1} $^{95m}TcO_4^-$ (see Figure 5.1A) or 3-40 kBq mL^{-1} $^{99m}TcO_4^-$ (see Figure 5.1B).

5.2.6 Analytical Pretreatment Procedures

At set intervals after addition of TcO_4^- to the nutrient solution, the plants were harvested and weighed. The activity concentrations of ^{95m}Tc or ^{99m}Tc in leaves, stem and roots, and used nutrient solution were determined. Both mass (activity) balances of ^{95m}Tc and ^{99m}Tc were 98 ± 1% (95% confidence limit). Before homogenisation of the leaves either $^{99m}TcO_4^-$ or ^{95m}TcX (for preparation see section 5.2.8) was added to the homogenisation buffer (see Figure 5.1) resulting in final concentrations of 6 kBq mL^{-1} or 1.0 kBq mL^{-1}, respectively. The leaves were homogenised in a buffer (70 mmol L^{-1} HEPES and 8 mmol L^{-1} NaCl; pH 7.0) using a Potter-Elvehjem homogeniser at 0-5°C; 3 g buffer was used per g fresh weight of leaves. Firstly, the ^{95m}Tc and ^{99m}Tc activity concentrations in the leaf homogenate were determined. Secondly, the homogenate was filtered through a 0.22-μm Millex-GV filter. In the filtrate the ^{95m}Tc and ^{99m}Tc activity concentrations were determined. Finally, the filtrates were analysed for TcO_4^- and TcX by SEC separation, and off-line γ-counting of ^{99m}Tc and ^{95m}Tc in fractionated aliquots.

5.2.7 Separation

The retention characteristics in size-exclusion chromatography are represented by the distribution coefficient $K_d = (V_e - V_0) / V_i$, where V_e refers to the elution volume of the solute.[15] The column recovery of Tc was 100 ± 1% (95% confidence limit). The chromatograms of the filtered homogenate of the leaves of spinach plants grown on TcO_4^- show two Tc species, i.e., TcX ($K_d = 1.0$) and TcO_4^- ($K_d = 3.0$). No shift in retention times of TcX with incubation time of plants was observed. Figure 5.2 shows typical chromatograms obtained for filtrates of spinach plant grown on $^{99}TcO_4^-$ containing nutrient solution spiked with either $^{95m}TcO_4^-$ (A) or $^{99m}TcO_4^-$ (C).

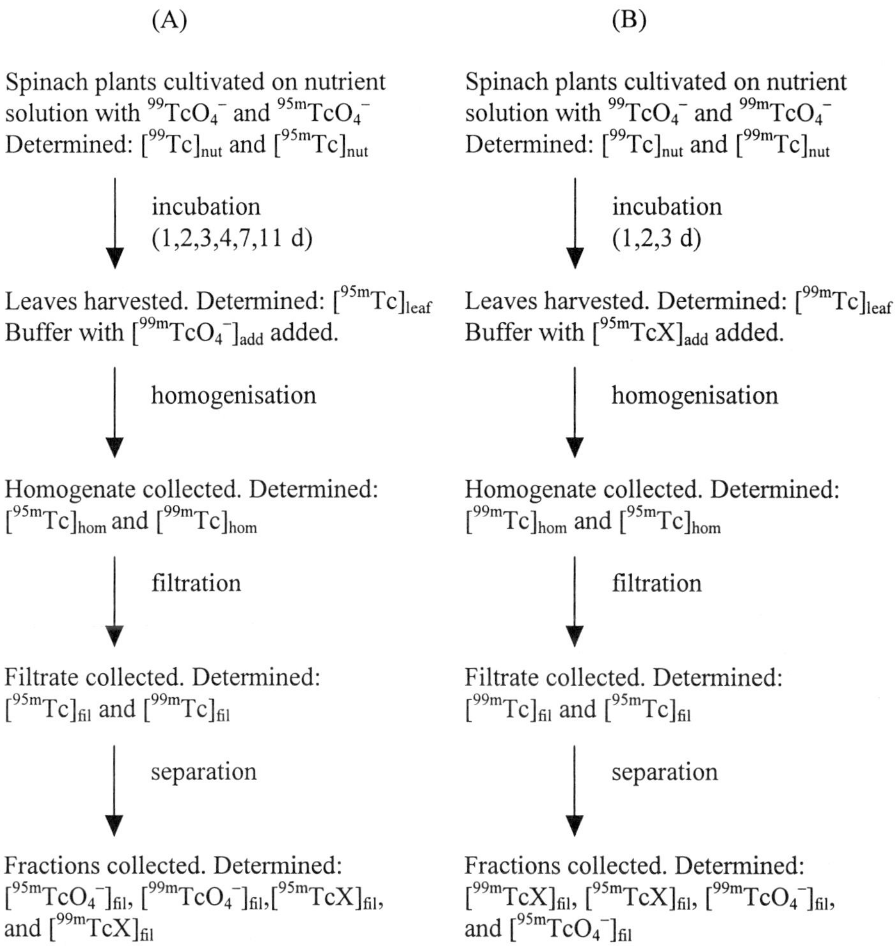

Figure 5.1. Flowchart of pretreatment and analytical procedures. [Tc], $[TcO_4^-]$, and [TcX] refer to the concentration (in mol g^{-1}) of total Tc, TcO_4^-, and TcX, respectively. The indices 95m, 99m and 99 refer to the nuclides. The subscripts nut, leaf, add, hom, and fil refer to the species concentration in the nutrient solution at the start of the incubation, in the leaves, in the combination of homogenisation buffer and leaves just before the homogenisation step, in the homogenate after the homogenisation step, and in the filtrate, respectively.

These tracing chromatograms are corrected for procedural artefacts by using the results from the complementary procedural control chromatograms (B and D), which were obtained by addition of either $^{99m}TcO_4^-$ (B) or ^{95m}TcX (D) during the analytical pretreatment. A detailed description of calculation procedures involved is given in the theory section.

5.2.8 Preparation of ^{95m}TcX Control Tracer

To synthesise the ^{95m}TcX control tracer, fifteen spinach plants were grown for 11 d on a nutrient solution containing 250 Bq mL^{-1} $^{95m}TcO_4^-$. All the leaves were pooled and homogenised in a buffer (70 mmol L^{-1} HEPES and 8 mmol L^{-1} NaCl; pH 7.0) (3 g buffer g^{-1} fresh weight), using a Potter-Elvehjem homogeniser at 0-5°C. The homogenate was centrifuged (20 min, 15,000 *g*, 4°C). The activity concentration of ^{95m}TcX in the collected supernatant was determined (93% of the total ^{95m}Tc present in the supernatant) and, subsequently, the supernatant was stored at –20°C, until use as ^{95m}TcX control tracer.

5.2.9 Detection

The detection of ^{99}Tc is described in section 3.2.7. ^{95m}Tc and ^{99m}Tc were measured, when necessary simultaneously, with a Wallac (Turku, Finland) 1480 Automatic 3" gamma counter, using appropriate energy windows settings (^{99m}Tc: 104-162 keV; ^{95m}Tc 162-240 keV).

5.3 THEORY

Preliminary experiments showed that soluble ^{99}Tc in spinach plants could be differentiated into two fractions, *viz.* $^{99}TcO_4^-$ and ^{99}TcX, i.e., reduced ^{99}Tc bound to unknown ligand(s). A method has been developed in which spinach plants are cultivated on substrate solutions containing $^{99}TcO_4^-$ and either $^{95m}TcO_4^-$ (see Figure 5.1A) or $^{99m}TcO_4^-$ (see Figure 5.1B) to allow easy radiotracing of ^{99}Tc. Furthermore, in the pretreatment procedure prior to fractionation another radiotracer is added to be able to distinguish between $^{99}TcO_4^-$ and ^{99}TcX correctly, i.e., taking into account procedural conversions (TcO_4^- to TcX and vice versa) and losses (adsorption to plant residue or filter material). To this end, in Figure 5.1A, $^{99m}TcO_4^-$ is added and in Figure 5.1B, ^{95m}TcX is added. A mathematical model of the analytical procedure is derived below.

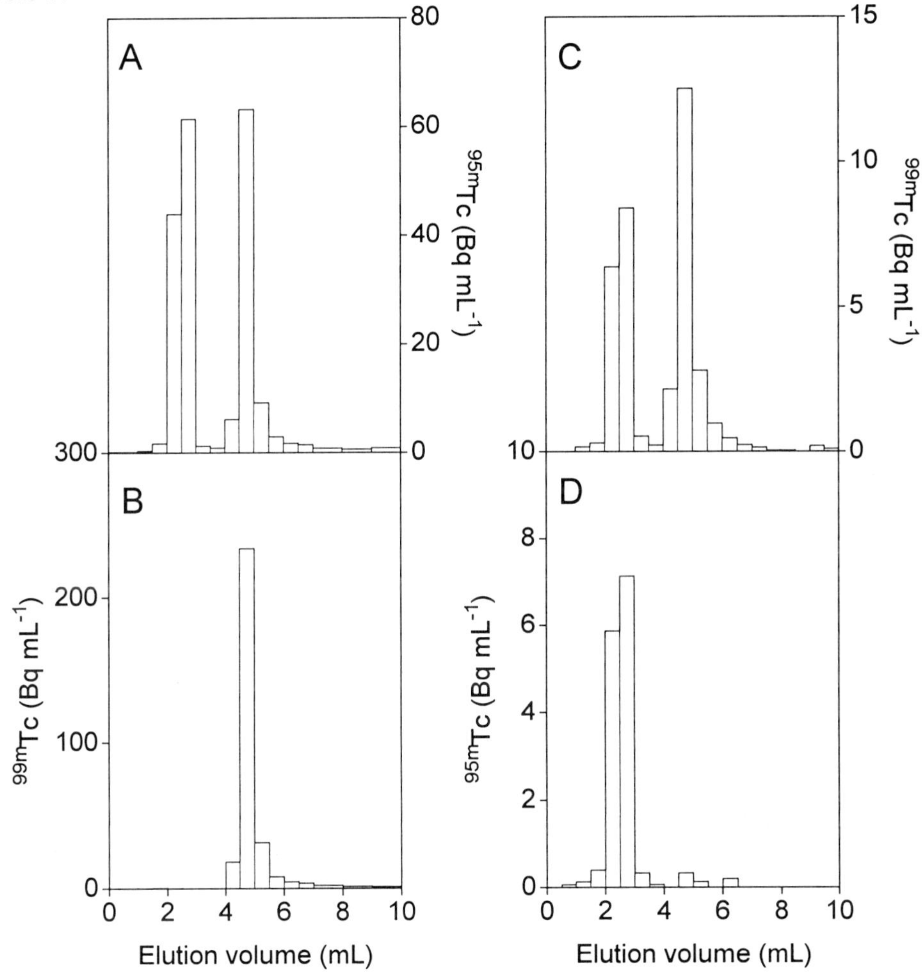

Figure 5.2. Separation of TcX (V_e: 2.7 mL) and TcO_4^- (V_e: 4.8 mL). Separation of the filtered homogenate of leaves of a spinach plant grown on $^{99}TcO_4^-$ traced with $^{95m}TcO_4^-$ (A), and monitored with $^{99m}TcO_4^-$ control tracer (B); spinach plant grown on $^{99}TcO_4^-$ traced with $^{99m}TcO_4^-$ (C), and monitored with ^{95m}TcX control tracer (D).

5.3.1 Tc Species Determination

In Figure 5.1 we assume that the activity concentration of ^{99}Tc in the leaves is given by

$$[^{99}\mathrm{Tc}]_{\mathrm{leaf}} = \frac{[^{99}\mathrm{Tc}]_{\mathrm{nut}}}{[^{95m}\mathrm{Tc}]_{\mathrm{nut}}}[^{95m}\mathrm{Tc}]_{\mathrm{leaf}} \qquad \text{(Figure 5.1A)} \qquad (5.1)$$

and

$$[^{99}\mathrm{Tc}]_{\mathrm{leaf}} = \frac{[^{99}\mathrm{Tc}]_{\mathrm{nut}}}{[^{99m}\mathrm{Tc}]_{\mathrm{nut}}}[^{99m}\mathrm{Tc}]_{\mathrm{leaf}} \qquad \text{(Figure 5.1B)} \qquad (5.2)$$

and

$$[^{99}\mathrm{Tc}]_{\mathrm{leaf}} = [^{99}\mathrm{TcO_4}^-]_{\mathrm{leaf}} + [^{99}\mathrm{TcX}]_{\mathrm{leaf}} \qquad (5.3)$$

From Figure 5.1A, $[^{99}\mathrm{TcO_4}^-]_{\mathrm{leaf}}$ can be derived:

$$[^{99}\mathrm{TcO_4}^-]_{\mathrm{leaf}} = \frac{[^{99}\mathrm{Tc}]_{\mathrm{nut}}}{[^{95m}\mathrm{Tc}]_{\mathrm{nut}}}[^{95m}\mathrm{TcO_4}^-]_{\mathrm{leaf}} \qquad (5.4)$$

If we assume partial conversion of $[^{99}\mathrm{TcO_4}^-]_{\mathrm{leaf}}$ to $[^{99}\mathrm{TcX}]_{\mathrm{leaf}}$ in the homogenisation procedure, then a ***procedural conversion factor*** C_1 for $\mathrm{TcO_4}^-$, can be introduced based on addition of $^{99m}\mathrm{TcO_4}^-$ to the homogenisation buffer:

$$[^{95m}\mathrm{TcO_4}^-]_{\mathrm{hom}} = C_1 \frac{[^{95m}\mathrm{Tc}]_{\mathrm{hom}}}{[^{95m}\mathrm{Tc}]_{\mathrm{leaf}}}[^{95m}\mathrm{TcO_4}^-]_{\mathrm{leaf}} \qquad (5.5)$$

where $\frac{[^{95m}\mathrm{Tc}]_{\mathrm{hom}}}{[^{95m}\mathrm{Tc}]_{\mathrm{leaf}}}$ denotes the dilution factor induced by addition of the homogenisation buffer, and

$$C_1 = \frac{[^{99m}\mathrm{TcO_4}^-]_{hom}}{[^{99m}\mathrm{TcO_4}^-]_{add}} \tag{5.6}$$

The filtration step might result in loss of TcO_4^- due to adsorption on filter and/or residual plant material. If we introduce a ***procedural recovery factor R_1*** for TcO_4^-, the following equations are obtained:

$$[^{95m}\mathrm{TcO_4}^-]_{fil} = \frac{m_{hom}}{m_{fil}} R_1 [^{95m}\mathrm{TcO_4}^-]_{hom} \tag{5.7}$$

and

$$[^{99m}\mathrm{TcO_4}^-]_{fil} = \frac{m_{hom}}{m_{fil}} R_1 [^{99m}\mathrm{TcO_4}^-]_{hom} \tag{5.8}$$

where m_{fil} refers to the mass of the filtrate (in g) and m_{hom} refers to the mass of the homogenate (in g). Substitution of eqs 5.5-8 in eq 5.4 yields the following expression for the calculation of $[^{99}\mathrm{TcO_4}^-]_{leaf}$:

$$[^{99}\mathrm{TcO_4}^-]_{leaf} = \frac{[^{99}\mathrm{Tc}]_{nut}}{[^{95m}\mathrm{Tc}]_{nut}} \frac{[^{99m}\mathrm{TcO_4}^-]_{add}}{[^{99m}\mathrm{TcO_4}^-]_{fil}} \frac{[^{95m}\mathrm{Tc}]_{leaf}}{[^{95m}\mathrm{Tc}]_{hom}} [^{95m}\mathrm{TcO_4}^-]_{fil} \tag{5.9}$$

From Figure 5.1B, $[^{99}\mathrm{TcX}]_{leaf}$ can be derived:

$$[^{99}\mathrm{TcX}]_{leaf} = \frac{[^{99}\mathrm{Tc}]_{nut}}{[^{99m}\mathrm{Tc}]_{nut}} [^{99m}\mathrm{TcX}]_{leaf} \tag{5.10}$$

If we assume partial conversion of $[^{99}\mathrm{TcX}]_{leaf}$ to $[^{99}\mathrm{TcO_4}^-]_{leaf}$ in the homogenisation procedure, then a ***procedural conversion factor C_2*** for TcX, can be introduced based on addition of ^{95m}TcX to the homogenisation buffer:

$$[^{99m}\mathrm{TcX}]_{hom} = C_2 \frac{[^{99m}\mathrm{Tc}]_{hom}}{[^{99m}\mathrm{Tc}]_{leaf}} [^{99m}\mathrm{TcX}]_{leaf} \tag{5.11}$$

where $\frac{[^{99m}\mathrm{Tc}]_{hom}}{[^{99m}\mathrm{Tc}]_{leaf}}$ denotes the dilution factor induced by addition of the homogenisation buffer, and

$$C_2 = \frac{[^{95m}\mathrm{TcX}]_{hom}}{[^{95m}\mathrm{TcX}]_{add}} \tag{5.12}$$

The filtration step might result in loss of TcX due to adsorption on filter and/or residual plant material. If we introduce a ***procedural recovery factor R_2*** for TcX, the following equations are obtained:

$$[^{99m}\mathrm{TcX}]_{fil} = \frac{m_{hom}}{m_{fil}} R_2 [^{99m}\mathrm{TcX}]_{hom} \tag{5.13}$$

and

$$[^{95m}\mathrm{TcX}]_{fil} = \frac{m_{hom}}{m_{fil}} R_2 [^{95m}\mathrm{TcX}]_{hom} \tag{5.14}$$

Substitution of eqs 5.11-14 in eq 5.10 yields the following expression for the calculation of $[^{99}\mathrm{TcX}]_{leaf}$:

$$[^{99}\mathrm{TcX}]_{leaf} = \frac{[^{99}\mathrm{Tc}]_{nut}}{[^{99m}\mathrm{Tc}]_{nut}} \frac{[^{95m}\mathrm{TcX}]_{add}}{[^{95m}\mathrm{TcX}]_{fil}} \frac{[^{99m}\mathrm{Tc}]_{leaf}}{[^{99m}\mathrm{Tc}]_{hom}} [^{99m}\mathrm{TcX}]_{fil} \tag{5.15}$$

5.3.2 Procedural Conversion and Recovery Factors

Although it is not necessary to know R_1, R_2, C_1, and C_2 for the Tc species determination, since they cancel out in eqs 5.9 and 5.15, it is important to know them to get more insight in procedural artefacts. R_1 and R_2 can be derived by substitution of eqs 5.7 and 5.14 in

$$[^{95m}Tc]_{hom} = [^{95m}TcO_4^-]_{hom} + [^{95m}TcX]_{hom} \tag{5.16}$$

and eqs 5.8 and 5.13 in

$$[^{99m}Tc]_{hom} = [^{99m}TcO_4^-]_{hom} + [^{99m}TcX]_{hom} \tag{5.17}$$

yielding

$$[^{95m}Tc]_{hom} = \left(\frac{[^{95m}TcO_4^-]_{fil}}{L_1} + \frac{[^{95m}TcX]_{fil}}{L_2}\right)\frac{m_{fil}}{m_{hom}} \tag{5.18}$$

and

$$[^{99m}Tc]_{hom} = \left(\frac{[^{99m}TcO_4^-]_{fil}}{L_1} + \frac{[^{99m}TcX]_{fil}}{L_2}\right)\frac{m_{fil}}{m_{hom}} \tag{5.19}$$

After solving the eqs 5.18 and 5.19, the following expressions are found for R_1 and R_2 (if R_1 and $R_2 \neq 0$).

$$R_1 = \frac{m_{fil}\left([^{95m}TcO_4^-]_{fil}[^{99m}TcX]_{fil} - [^{95m}TcX]_{fil}[^{99m}TcO_4^-]_{fil}\right)}{m_{hom}\left([^{95m}Tc]_{hom}[^{99m}TcX]_{fil} - [^{95m}TcX]_{fil}[^{99m}Tc]_{hom}\right)} \tag{5.20}$$

and

$$R_2 = \frac{m_{fil}\left([^{95m}TcO_4^-]_{fil}[^{99m}TcX]_{fil} - [^{95m}TcX]_{fil}[^{99m}TcO_4^-]_{fil}\right)}{m_{hom}\left([^{95m}TcO_4^-]_{fil}[^{99m}Tc]_{hom} - [^{95m}Tc]_{hcm}[^{99m}TcO_4^-]_{fil}\right)} \tag{5.21}$$

C_1 and C_2 can be derived easily now:

$$C_1 = \frac{m_{\mathrm{hom}}[^{99m}\mathrm{TcO_4}^-]_{\mathrm{fil}}}{m_{\mathrm{fil}} L_1 [^{99m}\mathrm{TcO_4}^-]_{\mathrm{add}}} \qquad (5.22)$$

and

$$C_2 = \frac{m_{\mathrm{hom}}[^{95m}\mathrm{TcX}]_{\mathrm{fil}}}{m_{\mathrm{fil}} L_2 [^{95m}\mathrm{TcX}]_{\mathrm{add}}} \qquad (5.23)$$

5.4 RESULTS AND DISCUSSION

Using a $^{95m}TcO_4^-$ or a $^{99m}TcO_4^-$ radiotracer during cultivation (see Figures 5.1A and B) it is possible to trace ^{99}Tc in the leaves by direct counting of the leaves. In this way an indication of the total ^{99}Tc concentration in plants in time is obtained. From Figure 5.3 it is clear that spinach plants are able to accumulate ^{99}Tc. Both radiotracers yield a similar trend for ^{99}Tc uptake, although ^{99m}Tc can only be followed for a short incubation time due to the short half-life. Since the ^{99}Tc concentration in the nutrient solution is 0.85 nmol g^{-1}, the concentration factor after 11 d is about 100 (g g^{-1}), what is in accordance with literature.[19]

Species information can only be derived indirectly after homogenisation of spinach plants although *in-situ* determination in the plant would be preferable. However, so far no non-destructive speciation techniques exist to be able to differentiate between Tc species on low levels. Before homogenisation either $^{99m}TcO_4^-$ or ^{95m}TcX (see Figures 5.1A and B) are added to correct for procedural conversions and losses, according to the theory developed above.

Firstly, procedural losses of TcO_4^- and TcX can be studied. As both species might have different affinities for plant residue and/or filter material, procedural loss factors R_1 and R_2 can be established for TcO_4^- and TcX, respectively. R_1 and R_2 are confined to values between 0 and 1, where 0 and 1 correlate with complete loss and complete recovery, respectively. Figure 5.4 relates these factors to incubation time. Surprisingly, TcO_4^- losses in the analytical procedure increase with cultivation time, what suggests that spinach plants may form high molecular weight compounds with a high affinity for

TcO_4^-, e.g., polysaccharides, such as polygalacturonates which are known to reduce and bind Fe^{III}, V^{V}, Mo^{VI}, and Cr^{VI}.[20] For TcX the recovery is nearly quantitative and independent of incubation time. It is clear that the selective adsorption of TcO_4^- observed on either the residue or the filter material, particularly at long incubation times, severely affects the results of uncorrected speciation data.

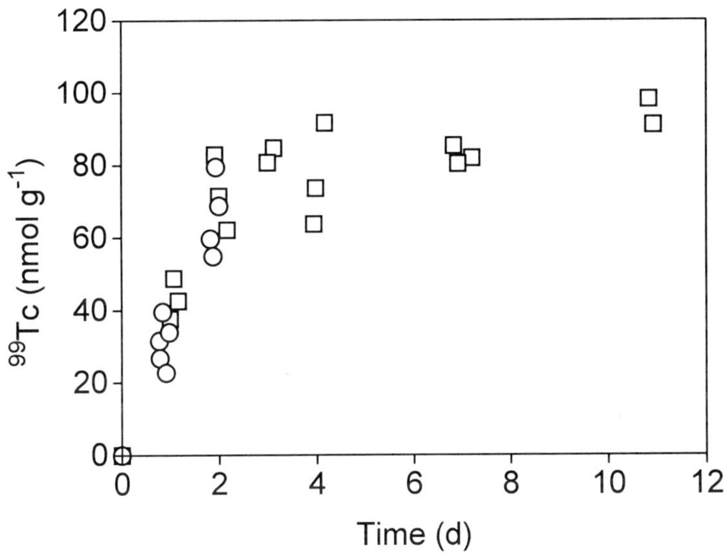

Figure 5.3. ^{99}Tc concentration in the leaves of spinach plants grown on $^{99}TcO_4^-$ and traced with either $^{95m}TcO_4^-$ (□) or $^{99m}TcO_4^-$ (O) as a function of incubation time.

Secondly, the transition between TcO_4^- and TcX during the homogenisation procedure can be studied. Since the $^{99m}TcO_4^-$ control tracer for investigation of the procedural reduction of TcO_4^- to TcX is not contaminated with other ^{99m}Tc species, the procedural conversion factor C_1 is confined to values between 0 and 1, where 0 and 1 correlate with complete reduction and no reduction, respectively. However, the ^{95m}TcX control tracer for investigation of procedural

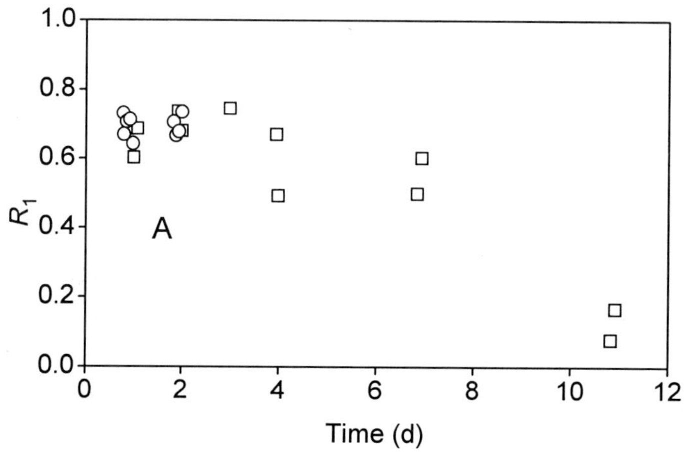

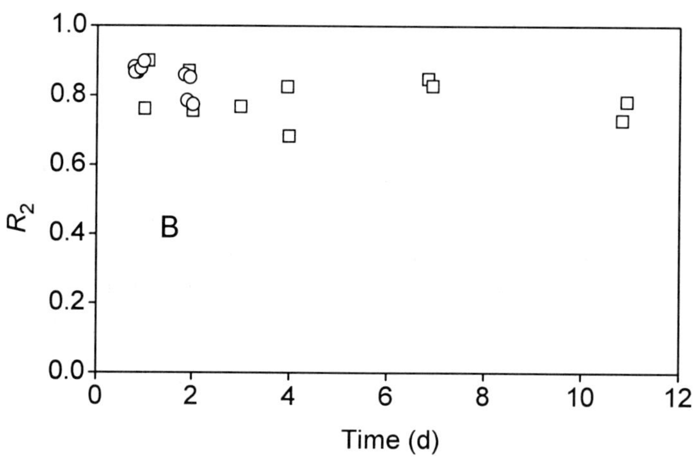

Figure 5.4. Procedural recovery as a function of incubation time. $^{99m}TcO_4^-$ control tracer (□) and ^{95m}TcX control tracer (○). (A) R_1 (procedural recovery correction factor for TcO_4^-). (B) R_2 (procedural recovery correction factor for TcX).

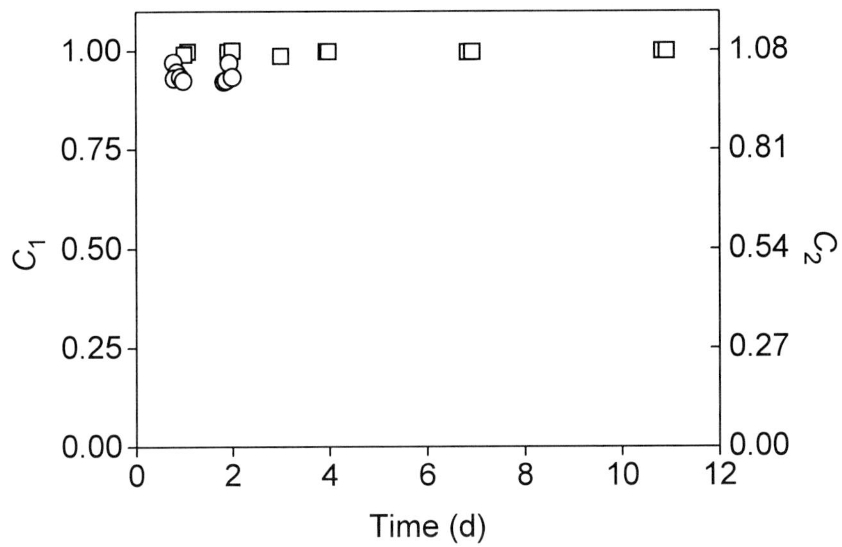

Figure 5.5. Procedural conversion of TcO_4^- and TcX as a function of incubation time. C_1 (□) (*left axis*) and C_2 (○) (*right axis*).

oxidation of TcX to TcO_4^- is not pure, but contains 7% of $^{95m}TcO_4^-$ (see experimental section), what confines the procedural conversion factor C_2 to values between 0 and 1.08, where 0, 1, and 1.08 correlate with complete oxidation of TcX, no conversion of TcX and TcO_4^-, and complete reduction of TcO_4^-, respectively. From Figure 5.5 it can be concluded that only a slight reduction might be expected since C_1 and C_2 range from 0.98 to 1.00 and from 1.00 to 1.05, respectively.

Finally, the concentrations of $^{99}TcO_4^-$ and ^{99}TcX in spinach leaves as a function of the incubation time are given in Figure 5.6. Since corrections are taken into account, this are the actual *in-situ* concentrations. $^{99}TcO_4^-$ and ^{99}TcX can be measured either directly or indirectly (Figures 5.6A and B), assuming only TcX in addition to TcO_4^- in spinach plants, by correlation of the "direct" data with the total ^{99}Tc activity concentrations from Figure 5.3. This has the advantage that TcX can also be determined in cases where the incubation time is larger than 2 d.

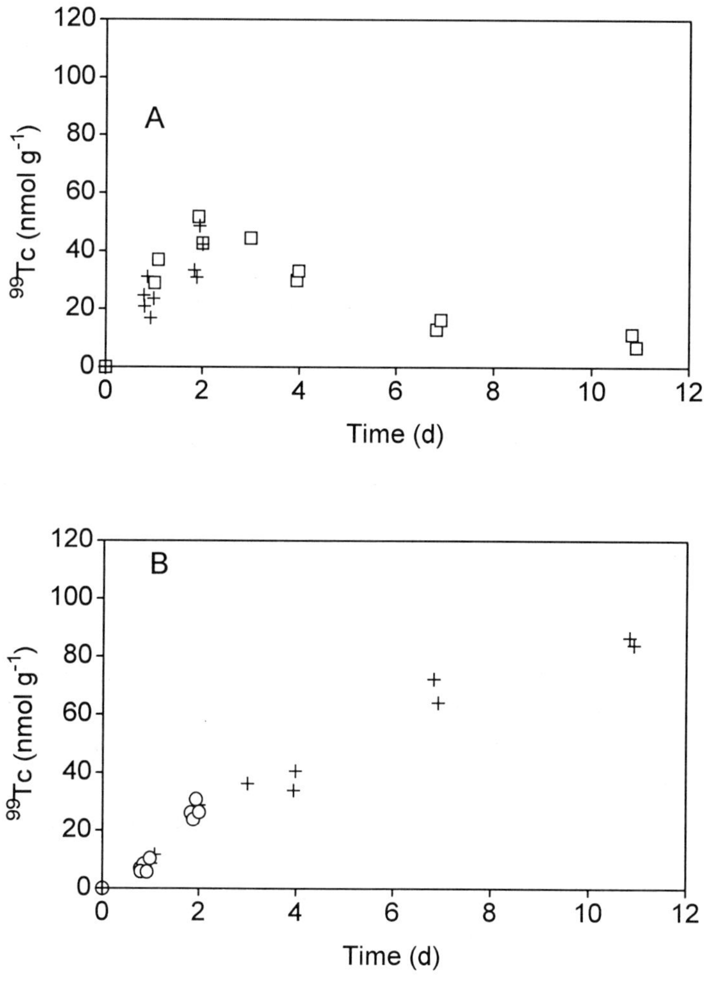

Figure 5.6. The $^{99}TcO_4^-$ and ^{99}TcX concentrations in the spinach leaves as function of incubation time. (A) $^{99}TcO_4^-$ concentration. (□) in leaves traced with ^{95m}Tc; (+) indirectly derived from ^{99}TcX concentration. (B) ^{99}TcX concentration. (O) in leaves traced with ^{99m}Tc; (+) indirectly derived from $^{99}TcO_4^-$ concentration.

5.5 CONCLUSIONS

This study illustrates a novel analytical speciation methodology, making use of a dual radiotracer. ^{99}Tc species ($^{99}TcO_4^-$ and plant-formed ^{99}TcX) could be traced by addition of $^{95m}TcO_4^-$ or $^{99m}TcO_4^-$ to the $^{99}TcO_4^-$ containing nutrient solution. Procedural artefacts (losses due to adsorption and transition of TcO_4^- and TcX) could be monitored by addition of $^{99m}TcO_4^-$ and ^{95m}TcX during the analytical pretreatment procedure. Major findings were significant procedural losses of $^{99}TcO_4^-$, increasing with incubation time, and insignificant redox conversions. Speciation of other elements, having suitable radioisotopes (e.g., Mn, Co, As, Se, Sn, Sb, and I), might benefit from this novel methodology.

5.6 REFERENCES

(1) Eiceman, G. A.; Hill, H. H., Jr.; Davani, B.; Gardea-Torresday, J. *Anal. Chem.* **1996**, *68*, 291R-308R.
(2) Pedersenbjergaard, S.; Greibokk, T. *J. High Resolut. Chromatogr.* **1996**, *19*, 597-607.
(3) Rothman, L. D. *Anal. Chem.* **1996**, *68*, 587R-598R.
(4) Dean, J. R.; Garden, L. M.; Armstrong, J.; Cresser, M. S.; Cave, M.; Watkins, P. *J. Anal. Atom. Spectrom.* **1997**, *12*, 19R-87R.
(5) Vela, N. P.; Olson, L. K.; Caruso, J. A. *Anal. Chem.* **1993**, *65*, 585A-597A.
(6) Burlingame, A. L.; Boyd, R. K.; Gaskell, S. J. *Anal. Chem.* **1996**, *68*, 599R-651R.
(7) Cornelis, R. *Analyst (Cambridge UK)* **1992**, *117*, 583-588.
(8) Holm, E. *Radiochim. Acta* **1993**, *63*, 57-62.
(9) Woodard-Blankenship, B.; Neel, J. W.; Papin, P. J. *Water, Air, Soil Pollut.* **1995**, *81*, 411-428.
(10) Shaw, G.; Bell, J. N. B. in *Plants and the chemical elements*; Farago, M. E., Ed.; VCH: Weinheim, 1994; Chapter 7.
(11) Krijger, G. C.; Claessens, H. A.; Wolterbeek, H. Th. *Chem. Spec. Bioav.* **1996**, *8*, 29-36.
(12) Lembrechts, J. F.; Desmet, G. M.; Overbeek, H. *Environ. Exp. Bot.* **1985**,

25, 355-360.
(13) Van Loon, L. R.; Desmet, G. M.; Cremers, A. *Health Phys.* **1989**, *57*, 309-314.
(14) Lembrechts, J. F.; Desmet, G. M. *Health Phys.* **1989**, *57*, 255-262.
(15) Garten, C. T., Jr.; Lomax, R. D. *Health Phys.* **1989**, *57*, 299-307.
(16) Harms, A. V.; Wolterbeek, H. Th.; Woroniecka, U. D. *Appl. Radiat. Isot.* **1996**, *47*, 605-610.
(17) Harms, A. V.; Van Elteren, J. T.; Claessens, H. A. *J. Chromatogr. A* **1996**, *755*, 219-225.
(18) Steiner, A. A. *Plant Soil* **1961**, *15*, 134-154.
(19) Desmet, G. M.; Dirkse, W. G. *Environ. Exp. Bot.* **1984**, *24*, 45-52.
(20) Deiana, S.; Dessi, A.; Micera, G.; Gessa, C.; de Cherchi, M. L. *Inorg. Chim. Acta* **1983**, *79*, 231-232.

Chapter 6

Characterisation of Technetium Species Induced in Spinach Plants: 1) Development of Separation Techniques*

6.1 INTRODUCTION

In Chapter 5 the chromatographic separation of soluble plant-induced TcX and TcO_4^- was performed with the method described in Chapter 4. This method, which is based on the hydrophobic interaction of TcO_4^- with the HEMA stationary phase,[1] was found unsuitable for differentiation between TcX species. Therefore, in this chapter both Sephadex low-pressure size-exclusion chromatography (SEC)[1] and high-performance reversed-phase ion-pair chromatography (RP-IPC)[2] are tested for their potential to differentiate between TcX species. As no TcX species have been identified so far, initially model compounds[1,3] are used to mimic these species. An anionic (Tc-DTPA) and a cationic (Tc-cyclam) compound, which were introduced in Chapter 3, are chosen for this purpose.[3] At a later stage, separation of TcX species induced in spinach plants is studied, based on the expertise obtained with the model compounds. Although high-performance RP-IPC[2,4] is supposed to give the best resolution, non-size effects in low-pressure SEC,[1] e.g., ion-exclusion,

* This chapter is a modified version of the paper "Search for Plant-Induced Technetium Species with Size-Exclusion and Reversed-Phase Ion-Pair Chromatography"; Harms, A. V.; Van Elteren J. T. *J. Radioanal. Nucl. Chem.* **1998**, *227*, in press.

hydrophobic interaction, and ion-exchange effects may enhance the separation of the compounds.[1] In particular, the parameters governing the separation are investigated for both chromatographic techniques, and it is attempted to relate chromatographic observations to structural information on TcX species.

6.2 EXPERIMENTAL

6.2.1 Reagents

All reagents used are described in section 3.2.1. ^{95m}Tc was obtained from Los Alamos National Laboratory (Los Alamos, NM, USA). The ^{95m}Tc stock solution also contained ^{97m}Tc (half-life 91 d), which decays to ^{97}Tc (half-life 2.6 × 10^6 y). The presence of these two isotopes presented no detection problems of ^{95m}Tc or ^{99m}Tc. The ratio of the amount of ^{95m}Tc and $^{97(m)}$Tc (in moles) in the ^{95m}Tc stock solution was time-dependent, i.e., approximately 0.48 exp (–0.34 t), where t refers to the time (in months) elapsed after receipt of the stock solution. Since the stock solution was used up to four months after receipt, the ^{95m}Tc/$^{97(m)}$Tc ratio was between 0.48 and 0.12.

6.2.2 Tc-DTPA complex

The synthesis of the Tc-DTPA complex is according to the procedure as described in section 3.2.3.

6.2.3 Tc-cyclam complex

The synthesis of the Tc-cyclam complex is according to the procedure as described in section 3.2.5.

6.2.4 Apparatus

The HPLC-apparatus used is described in section 4.2.4.

6.2.5 Reversed-Phase Ion-Pair Chromatography

A Macherey-Nagel MF-Plus (Metal-Free) Nucleosil C18 (150 mm × 4.6 mm i.d.; particle size: 5 μm; pore size: 100 Å) reversed-phase column (Alltech) and a MF-Plus Nucleosil C18 Guard column (10 mm × 4.6 mm i.d.) were used. Column specifications listed the dead volume V_m as 1.28 mL. The capacity factor k' is represented by $(V_e - V_m) / V_m$ or $V_e / V_m - 1$, where V_e refers to the elution volume of the solute. The used mobile phase consisted of a mixture of an aqueous solution of tetrabutylammonium (TBA) acetate (pH 5.9) and methanol.[5] A column temperature of 30°C was used.

6.2.6 Size-Exclusion Chromatography

The single-use PD-10 columns are described in section 3.2.2. All PD-10 columns were eluted with 12 mmol L^{-1} NaCl.

6.2.7 Preparation of Plant Induced TcX

Spinach plants were cultivated according to the procedure as described in section 5.2.4. After 10 d of growth, one batch of 15 spinach plants was placed in 100 mL fresh nutrient solution containing 250 Bq mL^{-1} $^{95m}TcO_4^-$. After 11 d incubation with TcO_4^- all plants were harvested. All leaves were pooled and homogenised in buffer (70 mmol L^{-1} HEPES and 8 mmol L^{-1} NaCl; pH 7.0), using a Potter-Elvehjem homogeniser at 0-5°C (3 g buffer g^{-1} fresh weight). The homogenate was centrifuged (15000 *g*, 4°C) for 20 minutes. The concentration of $^{95m}TcO_4^-$ in the collected supernatant, i.e., 7% of the total ^{95m}Tc present in the supernatant, was determined with the method described in chapter 5. Subsequently, the supernatant was stored at –20°C, until use.

Another batch of spinach plants was cultivated according to the procedure as described in section 5.2.4. After 10 d of growth, this second batch of spinach plants was weighed and placed in 100 mL fresh nutrient solution containing 250 Bq mL^{-1} $^{95m}TcO_4^-$. The concentration of ^{95m}Tc in the nutrient solution was 3.2 pmol L^{-1}. The total concentration of Tc (^{95m}Tc and $^{97(m)}Tc$) in the nutrient solution ranged between 7 and 27 pmol L^{-1} (see section 6.2.1). At set intervals after addition of $^{95m}TcO_4^-$ to the nutrient solution, the plants were harvested and weighed. The activity concentrations of ^{95m}Tc in leaves, stem and roots, and used nutrient solution were determined. The mass (activity) balance of ^{95m}Tc was 98 ± 2% (95% confidence limit). In Figure 6.1 it is shown that there was no significant difference in the accumulation of ^{95m}Tc between this batch of spinach plants (Figure 6.1A) and the batch described in section 5.2.5 (nutrient solution: 0.85 μmol L^{-1} $^{99}TcO_4^-$ and 250 Bq mL^{-1} $^{95m}TcO_4^-$; Figure 6.1B). These data suggest that the plant concentration factor (g mL^{-1}) is not dependent on the Tc concentration and that the uptake of Tc between these non-toxic concentration levels is a simple first-order process. These results are in agreement with those obtained by Van Loon et al.[6,7] and Hattink et al.,[8] who observed a similar linear relationship between the Tc concentrations in spinach and duckweed, respectively, and the concentration of Tc in the nutrient solution.

6.2.8 Detection

The detection of ^{95m}Tc, ^{99m}Tc, and ^{99}Tc proceeded as is described in sections 3.2.7 and 5.2.9.

6.3 RESULTS

6.3.1 Retention Behaviour of TcO_4^-

As separation characteristics on RP-IPC mainly depend on the organic modifier and ion-pair reagent concentrations,[9] both parameters were varied to study the retention behaviour of TcO_4^- (k') on the column. As can be seen from Figure 6.2 the mutual effect of a decreasing methanol concentration and an increasing TBA concentration leads to a increased retention of TcO_4^-. This can be explained by formation of an ion-pair between TcO_4^- and TBA and is in accordance with the behaviour of other oxo anions,[10] formed by elements such as Mo, Cr, and V. Mobile phase composition chosen for separation of TcO_4^-, Tc-DTPA, and Tc-cyclam is 1.8 mmol L^{-1} TBA acetate and 30% (v/v) methanol.

Separation by SEC (Sephadex G-25) can be regulated by changing the ionic strength of the mobile phase. The retention behaviour of TcO_4^- (K_d) heavily depends on the NaCl concentration; concentrations of 0, 5, and 12 mmol L^{-1} NaCl give K_d values of 0.6, 1.4, and 1.6, respectively.[1] This can be explained by the increase of hydrophobic interaction between TcO_4^- and the stationary phase.[11] Mobile phase composition chosen for separation of TcO_4^-, Tc-DTPA, and Tc-cyclam is 12 mmol L^{-1} NaCl.

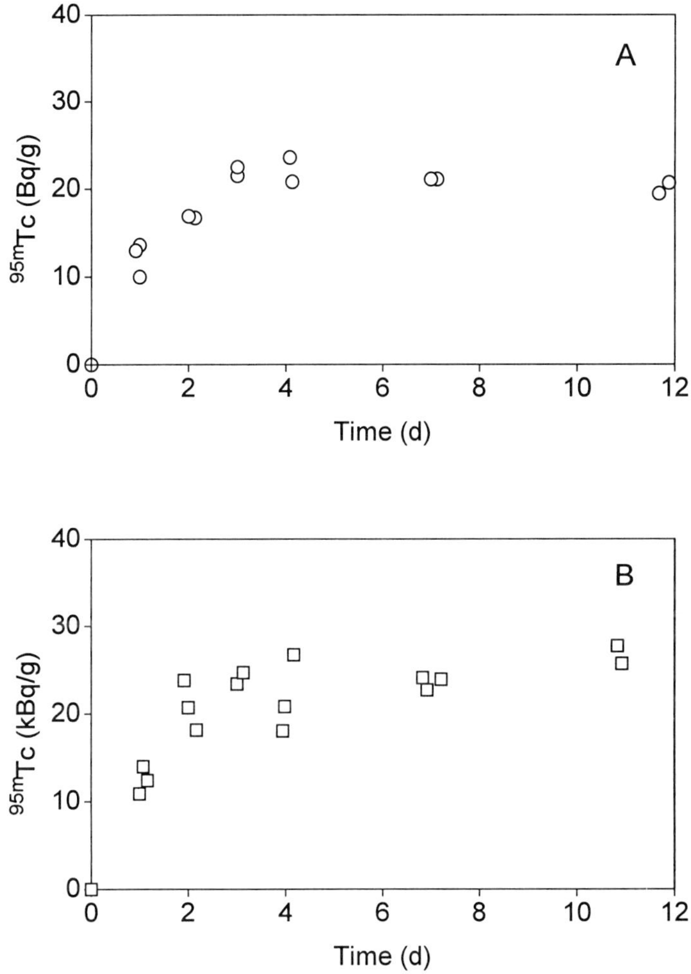

Figure 6.1. ^{95m}Tc concentration in the leaves of spinach plants as a function of the incubation time and Tc concentration in nutrient solution: (A) 250 Bq mL^{-1} $^{95m}TcO_4^-$ [3.2 pmol L^{-1} ^{95m}Tc]; (B) 250 Bq mL^{-1} $^{95m}TcO_4^-$ and 0.85 μmol L^{-1} $^{99}TcO_4^-$.

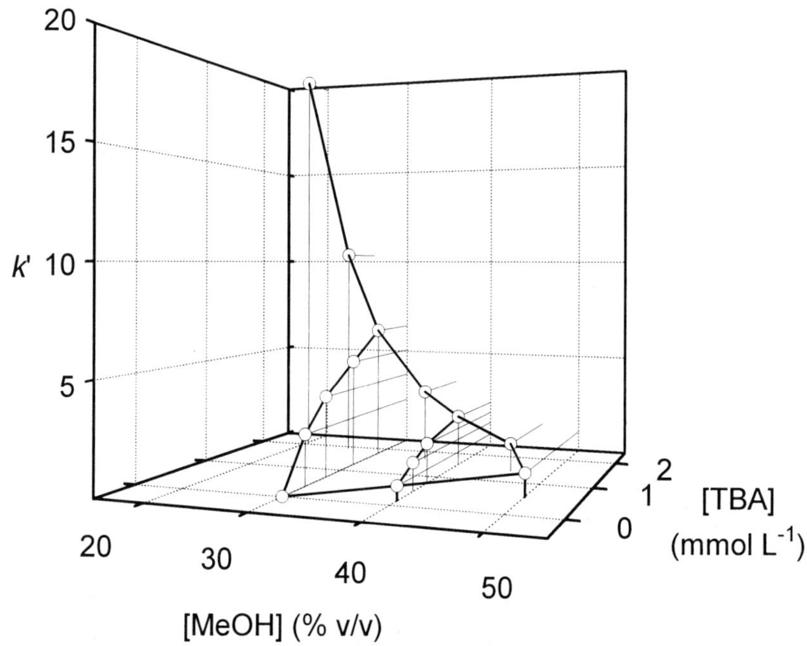

Figure 6.2. The capacity factor *k'* of $^{99}TcO_4^-$ as a function of the methanol and tetrabutylammonium acetate concentration in the mobile phase. Column: Nucleosil C18; column temperature: 30°C; flow rate: 1 mL min^{-1}.

6.3.2 Separation of TcO_4^-, Tc-DTPA, and Tc-cyclam

The chromatographic behaviour of TcO_4^-, Tc-DTPA, and Tc-cyclam on RP-IPC is shown in Figures 6.3A, B, and C, respectively. Although TBA will not form an ion-pair with Tc-cyclam, it is used to improve the retention of cationic metal chelates.[12] Only TcO_4^- (k' = 5.8) shows acceptable separation behaviour (narrow peak) and quantitative recovery (100%). Tc-cyclam and especially Tc-DTPA show strong interaction with the stationary phase, resulting in ill-defined peaks and recoveries of 85% and 11%, respectively, in the fractions between 0 and 25 mL. The reason for the tailing of the Tc-cyclam compound might be found in interaction of this cationic compound with residual silanol groups of the silica-based Nucleosil C18 stationary phase under the conditions chosen.[4,13] In this light, it is not understood why the recovery of an anionic compound such as Tc-DTPA is so low, unless dissociation[4] of this compound on the column is assumed. This seems a reasonable hypothesis since more than one peak is visible in the chromatogram.

The chromatographic behaviour of TcO_4^-, Tc-DTPA, and Tc-cyclam on SEC is shown in Figures 6.3D, E, and F, respectively. Since TcO_4^- has a low charge-density character, a K_d of 1.6 implies hydrophobic interaction with the Sephadex material.[1] The K_d of Tc-DTPA is 0.1, what is lower than expected for a solute of this size. The retention behaviour might be attributed to ion-exclusion due to repulsion between the anionic Tc-DTPA and slightly anionic Sephadex material.[1] The Tc-cyclam compound seems to be retained normally, resulting in a K_d of 0.6. For all three compounds quantitative recovery (>96% between 0 and 17 mL) is observed.

6.3.3 Investigation of TcX Characteristics

The chromatograms of TcX induced by spinach plants are shown in Figures 6.4A and B, for RP-IPC and SEC, respectively. The recovery of injected Tc is 61% and 96% for RP-IPC and SEC, respectively. Interference from residual TcO_4^- (7%) might effect the recovery of TcX slightly. In RP-IPC only one TcX compound can be observed ($k' = 3.2$). In SEC two subcompounds of TcX can be observed, TcX_1 ($K_d = 0.1$) and TcX_2 ($K_d = 1.1$), which consist of 28% and 68% of the injected Tc, respectively. Reinjection of TcX_1 (3.2-3.9 mL) and TcX_2 (9.8-10.5 mL) fractions eluted from the Sephadex material (Figure 6.4B) onto either RP-IPC or SEC give chromatograms as shown in Figures 6.5A to D. On Sephadex material TcX_1 ($K_d = 0.1$) and TcX_2 ($K_d = 1.1$) show recoveries of 84% and 96%, respectively. On Nucleosil C18 material TcX_1 and TcX_2 ($k' = 2.8$) show recoveries of 6 and 97%, respectively. When we speculate on the chemical properties of the subcompounds, TcX_1 might be related to high-molecular weight Tc-proteins, excluded on SEC due to size (>5 kDa) or anionic compounds, excluded on SEC due to charge. For TcX_2 non-size effects of SEC are observed, what assumes hydrophobic or cationic character. For both

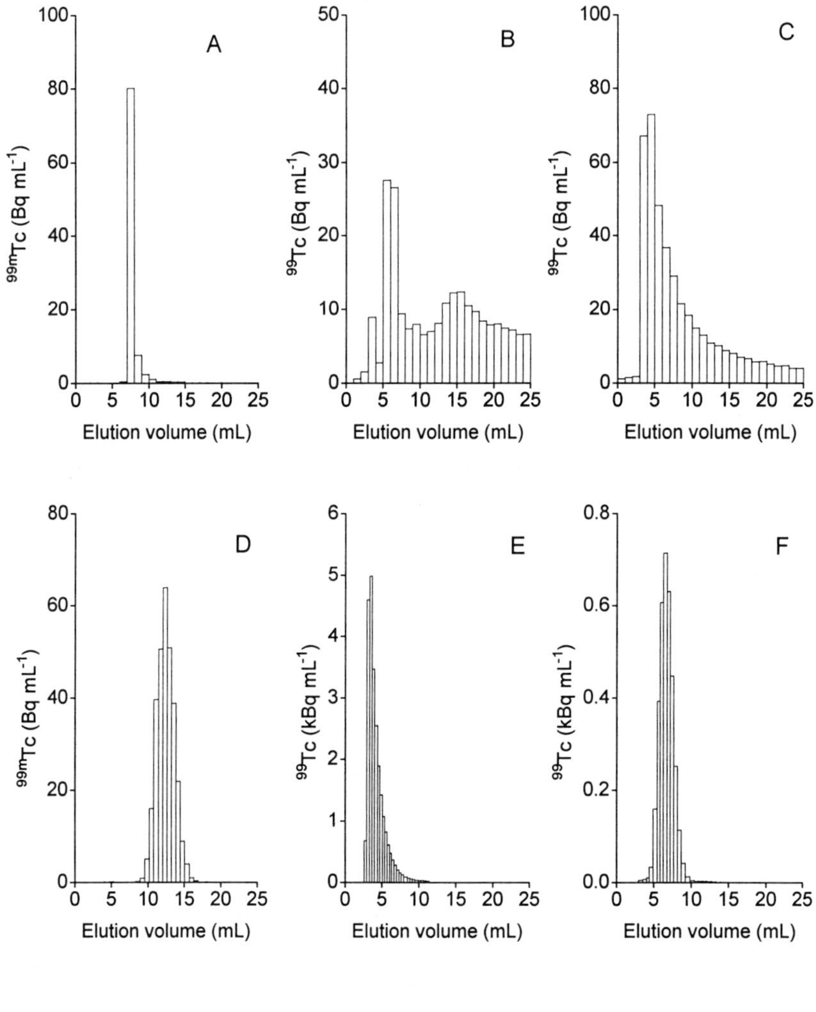

Figure 6.3. Retention behaviour of $^{99m}TcO_4^-$ (A), ^{99}Tc-DTPA (B), and ^{99}Tc-cyclam (C) on Nucleosil C18; mobile phase: 1.8 mmol L^{-1} TBA acetate (pH 5.9) and 30% (v/v) MeOH; column temperature: 30°C; flow rate: 1 mL min^{-1}. Retention behaviour of $^{99m}TcO_4^-$ (D), ^{99}Tc-DTPA (E), and ^{99}Tc-cyclam (F) on PD-10; mobile phase: 12 mmol L^{-1} NaCl; flow rate: gravity.

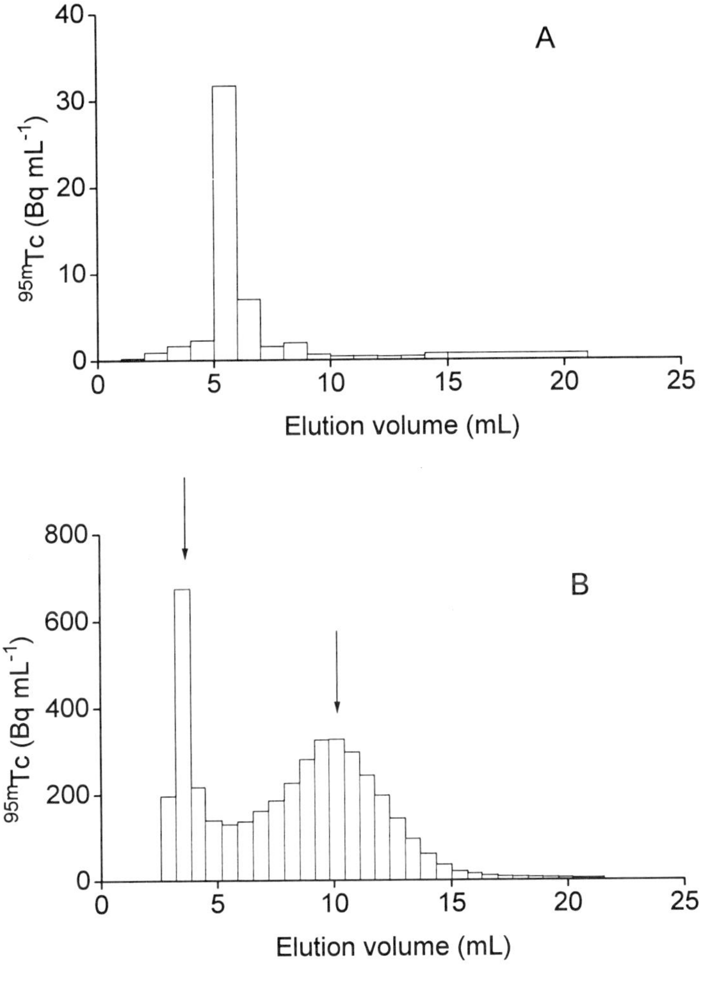

Figure 6.4. Separation of ^{95m}TcX. Column: Nucleosil C18 (A); mobile phase: 1.8 mmol L^{-1} TBA acetate (pH 5.9) and 30% (v/v) MeOH; column temperature: 30°C; flow rate: 1 mL min^{-1}. Column: PD-10 (B); mobile phase: 12 mmol L^{-1} NaCl; flow rate: gravity.

(9) Goldberg, A. P.; Nowakowska, E.; Antle, P. E.; Snyder, L. R. *J. Chromatogr.* **1984**, *316*, 241.
(10) Zhang, X.-S.; Zhu, X.-P.; Lin, C.-S. *Talanta* **1986**, *33*, 838.
(11) Verbeke, K.; Verbruggen, A. *J. Pharm. Biomed. Anal.* **1996**, *14*, 1209-1213.
(12) Miura, J. *Fresenius J. Anal. Chem.* **1992**, *344*, 247.
(13) Timerbaev, A. R.; Tsoi, I. G.; Petrukhin O. M. *Anal. Chim. Acta* **1992**, *269,* 229-238.

Chapter 7

Characterisation of Technetium Species Induced in Spinach Plants: 2) Combination of Selective Extraction and Chromatographic Techniques*

7.1 INTRODUCTION

To get insight into the nature and properties of TcX, it seems logic to use chromatographic separation techniques after homogenisation of the plant material. Generally, size-exclusion chromatography (SEC) is the method of choice for differentiation between TcO_4^- and sub-species of TcX, because of its "mild" separation characteristics.[1-6] However, due to poor resolution and non-size effects in SEC,[7] it is difficult to correlate the retention behaviour to molecular size. Although the information retrieved with reversed-phase ion-pair chromatography (RP-IPC) is generally more detailed as a result of the separation mechanisms involved, problems were encountered during separation of Tc species present in spinach leaf homogenate (see Chapter 6).[8] These problems might be related to irreversible retention of Tc species on the stationary phase. However, recent trials have shown, that a different type of

* This chapter is a slightly modified version of the paper "Characterisation of Technetium Species induced in Spinach Plants using a Combination of Selective Extraction and Chromatographic Techniques"; Harms, A. V.; Van Elteren, J. T. **1998**, in preparation.

RP-IPC has potential when combined with selective extraction of Tc-protein prior to separation. Several photosynthetically formed Tc^{V} species have been suggested to occur in plants[1,2] but to date none of them has been unequivocally identified. The Tc^{V}-cysteine complex is thought to be the major Tc^{V} species,[2,9-11] due to the affinity of Tc^{V} for N- and S-donor ligands.

In this chapter, the features of an extraction scheme are exploited to separate sub-species of TcX, induced by spinach plants grown on TcO_4^--containing nutrient solutions into classes of compounds. To this end, Tc^{V}-cysteine was synthesised according to a slightly modified literature procedure[12,13] and compared with sub-species of TcX. Both SEC and RP-IPC were used as separation methods for classification of individual species.[14-16] Combination of the crude differentiation by extraction with the more specific chromatographic identification of certain Tc species might yield a new insight into metabolic pathways of TcO_4^- transformation in plants.

7.2 EXPERIMENTAL

7.2.1 Reagents

All reagents used are described in sections 3.2.1 and 5.2.1.

7.2.2 Apparatus

The HPLC-apparatus used is described in section 4.2.4.

7.2.3 Cultivation of Plants

The spinach plants were cultivated according to the procedure as described in section 5.2.4.

7.2.4 Incubation of plants with TcO_4^-

After 10 d of growth, 15 spinach plants were placed in 100 mL fresh nutrient solution containing either 250 Bq mL^{-1} $^{95m}TcO_4^-$ (A) or 53 Bq mL^{-1} $^{99}TcO_4^-$ and 250 Bq mL^{-1} $^{95m}TcO_4^-$ (B). This corresponds to concentrations in the nutrient solutions of 7-27 pmol L^{-1} Tc (A) or 0.85 μmol L^{-1} Tc (B). In Chapter 6 it was demonstrated that these nutrient solutions gave transfer factors for Tc (g mL^{-1}), which were not significantly different. The assumption made in

Chapter 5, that both non-toxic Tc concentrations yielded similar distribution patterns of Tc species in spinach leaves, was confirmed with SEC. After 11 d all plants were harvested. All leaves were pooled and homogenised in buffer (70 mmol L^{-1} HEPES and 8 mmol L^{-1} NaCl; pH 7.0), using a Potter-Elvehjem homogeniser at 0-5°C (3 g buffer g^{-1} fresh weight). Subsequently, the leaf homogenate was stored at –20°C, until use.

7.2.5 Extraction

Several extraction techniques[5,17-19] were used to differentiate between classes of technetium species in the homogenate of spinach plants (see Figure 7.1). For quantification purposes fractions were collected and counted as described in section 7.2.9.

7.2.5.1 The homogenate was centrifuged (20 min, 15,000 *g*, 4°C). The amount of Tc in the pellet was determined. The pellet was resuspended in 1 mol L^{-1} sodium hydroxide and this suspension was centrifuged.

7.2.5.2 To precipitate hydrophobic Tc-proteins, ammonium sulphate (final concentration 2.1 mol L^{-1}) was added to the homogenate.[20] After centrifugation, acetone (final concentration 80% (w/w)) was added to the supernatant. The pellet was resuspended in water. Both the acetone-supernatant mixture and the resuspended pellet were centrifuged.

7.2.5.3 To precipitate hydrophilic Tc-proteins, acetone (final concentration 80% (w/w)) was added to the homogenate. The resulting precipitate was centrifuged.

7.2.5.4 Non-polar Tc species in the homogenate were extracted with diethyl ether. After 30 minutes of shaking the phases were separated.

7.2.5.5 The homogenate was shaked with a solution of 8-hydroxyquinoline 4% (w/w) in chloroform[21] to extract labile Tc^{V} species. Again, after 30 minutes of shaking the phases were separated.

7.2.6 Size-Exclusion Chromatography

A Pharmacia Fast Desalting HR 10/10 column filled with Sephadex G-25 Superfine grade material (100 mm × 10 mm i.d.; particle size range (wet): 17-69 μm; exclusion limit: 5 x 10^3 Dalton), operated at 20°C, was used. Blue Dextran 2000 was used to determine the void volume V_0 (2.1 mL) of this column. The specifications of the manufacturer listed an internal volume V_i of

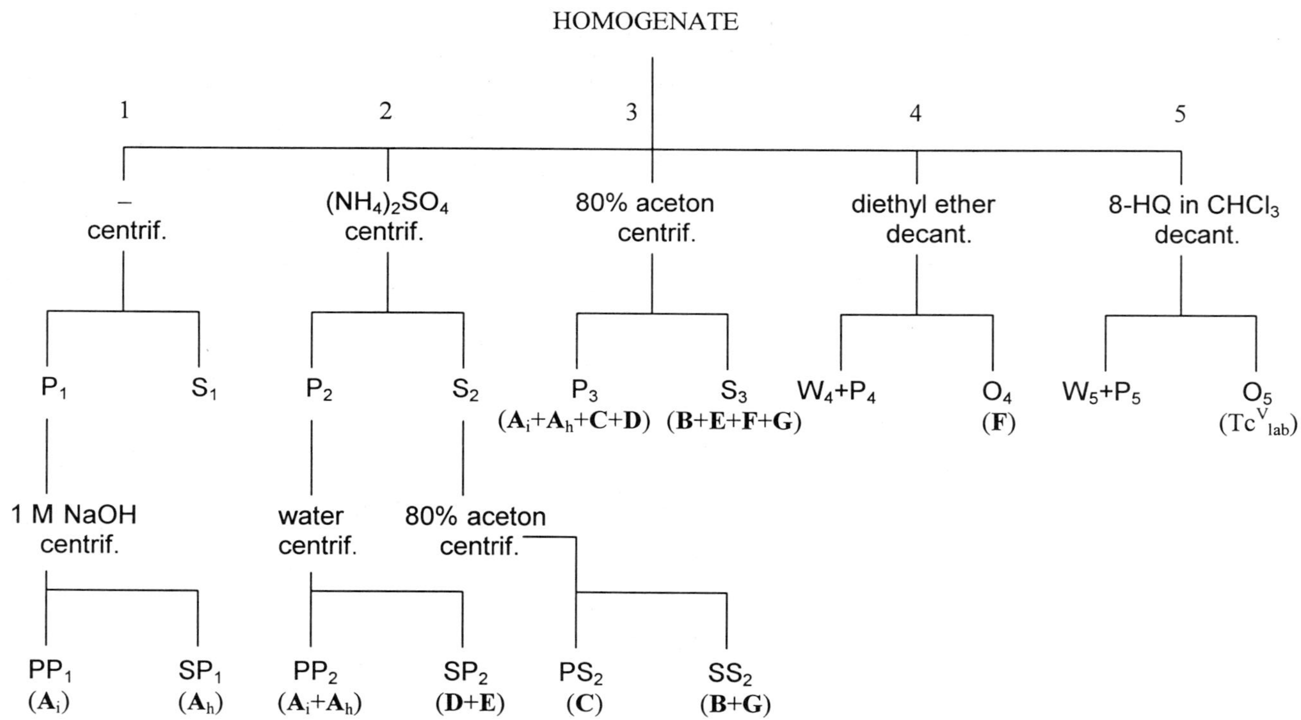

Figure 7.1. Extraction scheme. P, S, W, and O denote pellet, supernatant, aqueous phase, and organic phase, respectively. Double letters indicate an additional extraction step resulting in a new pellet and supernatant. Bold letters represent classes of compounds separated (see Table 7.1).

4.4 mL. For all separations a mobile phase consisting of 0.5 mol L^{-1} $CaCl_2$, at a flow rate of 1 mL min^{-1}, was used. Leaf homogenate was centrifuged; the resulting supernatant was injected and fractions were collected.

7.2.7 Reversed-Phase Ion-Pair Chromatography

The Nucleosil C18 column used is described in section 6.2.5. For all separations in this chapter a mobile phase consisting of a mixture of an aqueous solution of 0.05% (v/v) trifluoroacetic acid (TFA) (pH 2.2) and methanol, at a flow rate of 1 mL min^{-1}, was used. Gradients of 0 to 10% (v/v) methanol between 0 and 20 min and 10 to 30% (v/v) methanol between 20 and 30 min were used. The column was operated at 30°C. Leaf homogenate was subjected to precipitation with $(NH_4)_2SO_4$ (final concentration 2.1 mol L^{-1}), followed by centrifugation; the resulting supernatant was injected and fractions were collected.

7.2.8 Synthesis of Tc Complexes

The Tc^{IV}-cysteine and Tc^{V}-cysteine complexes were obtained by reduction of TcO_4^- and subsequent complexation by excess cysteine at low pH.[12,13] No conventional TcO_4^- reduction methods utilising Sn^{II} species was applied, because residual Sn^{II} might retain on the column, causing artefacts in subsequent chromatographic runs.[22,23]

To a 0.65 mL aliquot of 86 MBq L^{-1} (1.4 mmol L^{-1}) $^{99}TcO_4^-$ in 10 mmol L^{-1} trifluoroacetic acid (TFA) and 4 mmol L^{-1} NH_4OH was added 18 mg (0.11 mmol) L-cysteine HCl. The solution (pH 1.0) was heated for 5 minutes in a water bath at 55°C. The solution turned yellow (Tc^{V}-cysteine) immediately, and red (Tc^{IV}-cysteine) after 20 h of storage at 20°C.

The two sub-species could be separated with SEC what made it possible to follow the reduction (Figure 7.2). As can be seen, the TcO_4^- concentration decreases rapidly, leading to the formation of "yellow" Tc^{V}-cysteine in the first instance and "red" Tc^{IV}-cysteine in the second instance. Since we need the "yellow" Tc^{V}-cysteine for check purposes, it was decided to use the reaction mixture at 1 h incubation time to be sure that the amount of "red" Tc^{IV}-cysteine was negligible.

The Tc^{V}-glutathione complex was obtained by a similar type of synthesis. To a 0.33 mL aliquot of 86 MBq L^{-1} (1.4 mmol L^{-1}) $^{99}TcO_4^-$ in 10 mmol L^{-1}

trifluoroacetic acid (TFA) and 4 mmol L^{-1} NH_4OH was added 17 mg (55 μmol) glutathione (GSH). The solution (pH 3.0) was heated for 5 minutes in a water bath at 55°C. Subsequently, the solution turned yellow, but slower than for the Tc^V-cysteine complex.

Furthermore, to check for a procedural artefact due to pre- or on-column formation of a Tc^V-TFA complex, it was attempted to synthesise a Tc^V-TFA complex by the reaction of $Tc^VOCl_4^-$ and TFA in MeOH. To this end, 34 mg TFA was added to 8.0 g MeOH. Subsequently, 0.9 mg $(n\text{-}Bu)_4NTc^VOCl_4$, which was synthesised according to a previous published procedure,[24] was added to the TFA-MeOH solution. The resulting solution was analysed with the RP-IPC system described in section 7.2.5. The chromatogram showed one peak with excessive tailing with an identical elution volume as TcO_4^- (2.4 mL). This result might be explained by disproportionation or dissociation and subsequent oxidation of a labile Tc^V-TFA complex or Tc^VOCl_4.[25] In any case, this result indicates that no procedural artefact, due to pre- or on-column formation of a Tc^V-TFA complex, is to be expected.

7.2.9 Detection

The detection of ^{95m}Tc and ^{99}Tc proceeded as described in sections 5.2.9 and 3.2.7.

7.3 RESULTS AND DISCUSSION

7.3.1 Extraction

To get a separation of Tc species in classes, leaf homogenate was extracted as described in section 7.2.5.

7.3.1.1 After centrifugation of the homogenate, 83% of the total Tc in the homogenate remained in supernatant S_1. The remaining 17% Tc was found in the pellet P_1 (species $A_i + A_h$). This percentage is lower than obtained for maize (31%),[26,27] soybean (39%),[26] and maple tree leaves (78%),[5] but comparable with the percentage obtained for tomato leaves (16%).[28] Almost all Tc species in pellet P_1 (species A_h; 16% of the total Tc) can be extracted into a sodium hydroxide solution, which suggests binding of Tc to hemicellulose.

Hemicellulose refers to various poorly defined polysaccharides of the cell wall, excluding cellulose, which are soluble in dilute alkaline solution.[5]

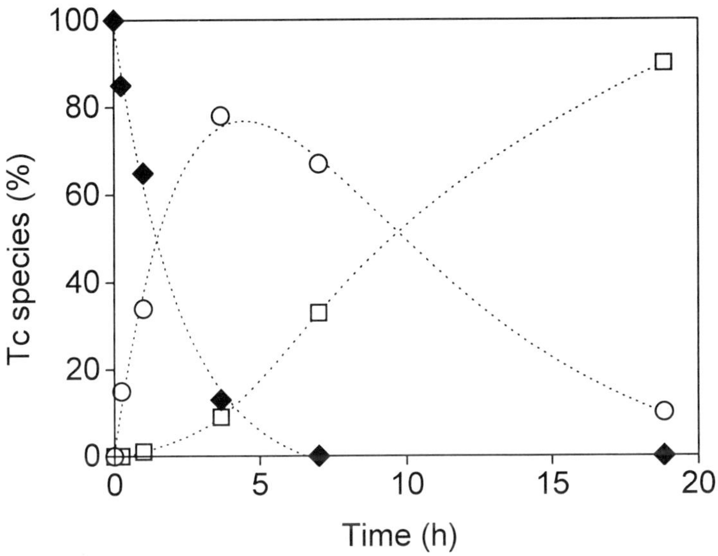

Figure 7.2. The reduction of TcO_4^- (◆) with L-cysteine at pH 1.0 resulting in the formation of Tc^V-cysteine (O) and Tc^{IV}-cysteine (□) complexes as function of incubation time. Incubation temperature: 20°C. Separation conditions: Sephadex Fast Desalting HR 10/10 column, mobile phase: 0.5 mol L^{-1} $CaCl_2$, flow rate: 1 mL min^{-1}.

7.3.1.2 Addition of ammonium sulphate to the homogenate and subsequent extraction of supernatant S_2 with 80% (w/w) acetone yielded 32% soluble Tc species. Representatives are TcO_4^- (species G) and/or non-protein Tc species (species B). Tc species which are soluble in ammonium sulphate solution, but subsequently precipitate in 80% (w/w) acetone (25% of the total Tc) probably consist of highly polar non-protein Tc species (species C). Tc species which precipitate in ammonium sulphate, but subsequently are soluble in water probably consist of Tc bound to hydrophilic proteins (species D) and/or hydrophobic proteins (species E).

7.3.1.3 Direct addition of acetone to the homogenate (final concentration 80% (w/w)) resulted in solubilisation of 42% of the total Tc in the homogenate. Tc species in this soluble fraction can be attributed to TcO_4^-, which is soluble in acetone,[29] non-protein Tc species (species B), and/or Tc bound to hydrophobic proteins (species E).

7.3.1.4 Very hydrophobic Tc species (species F) were not observed as the extraction with diethyl ether showed no radioactivity in the organic phase.

7.3.1.5 The presence of labile Tc^V species in the homogenate was tested with the extraction method developed by Hwang.[21] This method is based on the formation of a stable, uncharged complex of 8-hydroxyquinoline with Tc^V, which is extractable in chloroform. However, 99% of the Tc remained in the aqueous phase, indicating the virtual absence of labile Tc^V species, which is in accordance with the results obtained by Lembrechts et al..[2]

TcO_4^- was determined with column separation experiments (see section 7.3.2). In Table 7.1 the classes of Tc species are summarised.

Table 7.1. Classes of Tc species in the homogenate of spinach plant leaves

(A_h)	16%	Tc species found in the pellet soluble in 1 mol L^{-1} NaOH
(A_i)	1%	Tc species found in the pellet insoluble in 1 mol L^{-1} NaOH
(B)	25%	non-protein Tc species soluble in 80% (w/w) acetone
(C)	25%	highly polar non-protein Tc species
(D)	16%	Tc bound to hydrophilic protein
(E)	10%	Tc bound to hydrophobic protein
(F)	0%	Very hydrophobic Tc species
(G)	7%	TcO_4^-

7.3.2 Column Separation

The size-exclusion chromatogram (Figure 7.3A) of supernatant S_1 (see Figure 7.1) shows at least five Tc species eluting with elution volumes of 2.6, 5.0, 7.2, 10.3, and 13.2 mL, respectively. At 2.6 mL, species D and E are probably excluded due to size (MW > 5 kDa). Since a distribution coefficient

K_d of unity correlates with an elution volume of 6.5 mL, the later eluting species are retarded due to adsorptive forces, i.e., non-size effects, between the solute and column material. The last eluting species (V_e = 13.2 mL) showed an identical elution volume as TcO_4^-. The column recovery of Tc was 90%, suggesting that labile Tc species in the homogenate were adsorbed on the column.

The reversed-phase ion-pair chromatogram (Figure 7.3B) of supernatant S_2 (see Figure 7.1) shows at least three Tc species eluting with elution volumes of 2.4, 8.4, and 14.4 mL, respectively. The first eluting species (V_e = 2.4 mL) showed an identical elution volume as TcO_4^-. The column recovery of Tc was only 80%, but this is a low estimate since residual Tc still elutes from the column at an elution volume of 30 mL. This might be due to interaction of labile Tc species with the stationary phase.

The "yellow" Tc^V-cysteine complex after 1 h of incubation (see section 7.2.8) was separated with SEC (Figure 7.3C) and RP-IPC (Figure 7.3D) as well. In the size-exclusion chromatogram "red" Tc^{IV}-cysteine complex, "yellow" Tc^V-cysteine complex, and pertechnetate are observed at elution volumes of 6.5 (K_d = 1.0), 10.6 (K_d = 1.9), and 13.2 mL (K_d = 2.5), respectively. In the reversed-phase ion-pair chromatogram "red" Tc^{IV}-cysteine complex, pertechnetate, and "yellow" Tc^V-cysteine complex are observed at elution volumes of 1.8 mL (k' = 0.4), 2.4 mL (k' = 0.9), and 7.9 mL (k' = 5.2), respectively. In both SEC and RP-IPC the column recovery was quantitative. It is obvious that in both cases the amount of "red" Tc^{IV}-cysteine is negligible compared to "yellow" Tc^V-cysteine. Comparison of the Tc^V-cysteine chromatograms with those of the spinach plants shows that approximately 37% and 43% of the Tc injected can be attributed to Tc^V-cysteine in supernatant S_1 (Figure 7.3A; V_e = 10.3 mL) and S_2 (Figure 7.3B; V_e = 8.4 mL), respectively. Combining this with the extraction data from Figure 7.1, it can be deduced that approximately 25% of Tc in spinach leaf homogenate corresponds to Tc^V-cysteine and probably resembles species B in Table 7.1. It is not clear, whether this complex is anionic, e.g., $[Tc^VO(cys)_2]^-$ (H_2cys = L-cysteine), or neutral like the related $[Tc^VO(pen)(Hpen)]$ complex (H_2pen = D-penicillamine or β,β-dimethyl-cysteine),[30,31] although a structural resemblance with this latter complex, in which the penicillamine dianion is tridentate (*N*, *S*, *O*) and the penicillamine anion is bidentate (*N*, *S*),[30] is probable. Moreover, it is interesting

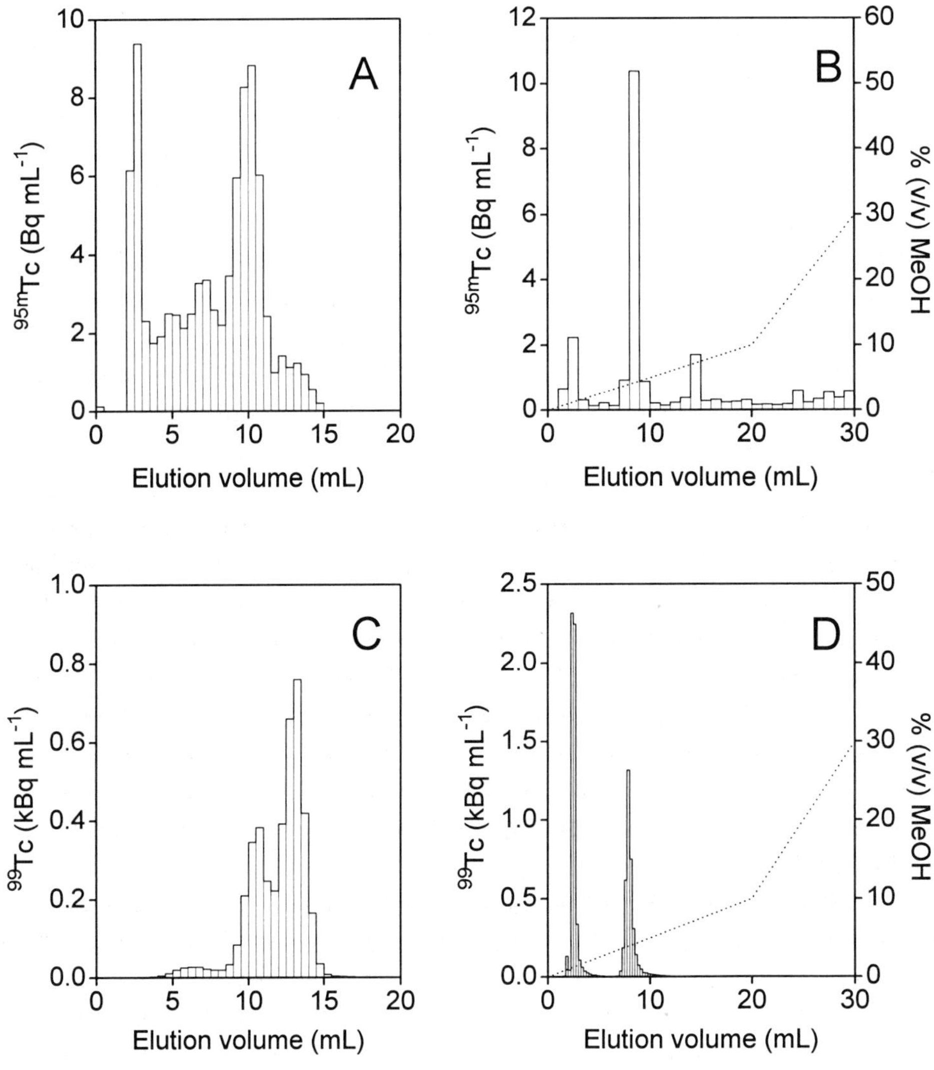

Figure 7.3. Chromatograms of supernatant S_1 (A), the supernatant S_2 (B), and Tc^V-cysteine (C and D) obtained with SEC (A and C) and RP-IPC (B and D). Column: Sephadex Fast Desalting HR 10/10 (A and C), column temperature: 20°C, mobile phase: 0.5 mol L^{-1} $CaCl_2$, flow rate: 1 mL min^{-1}. Column: Nucleosil C18 (B and D); column temperature: 30°C, mobile phase: 6.5 mmol L^{-1} TFA with MeOH gradient, flow rate: 1 mL min^{-1}. The dotted line in B and D describes the MeOH gradient.

that the structure of Tc^{V}O-EC[32] (EC = *N,N'*-1,2-ethylene-bis-L-cysteine) is thought to resemble the penicillamine complex.

It was suggested,[4,33] that Tc, like Se, can act as an analogue for S in S-containing amino acids, e.g., cysteine. In higher plants, SO_4^{2-} is reduced via bound GSH reductase, followed by incorporation of reduced S^{-II} into cysteine by cysteine synthetase. These reactions are carried out in the chloroplast using reductants from the photosynthesis.[34,35] The reduction of TcO_4^- is also thought to take place in the chloroplast. However, as was pointed out by Lembrechts,[36] it is unlikely that Tc can act as a S-analogue in thiols, because of the absence of the –II oxidation state. It seems more likely that the observed relation between Tc and S in these studies can be explained by the formation of a Tc-cysteine complex.

If we speculate on the nature of the other Tc species, other cysteine-moiety containing species, peptides such as the tripeptide glutathione (γ-L-Glu-L-Cys-Gly),[37,38] phytochelatines,[39] and proteins, are potential candidates as Tc-binding ligands. In fact, the chromatographic behaviour of Tc^{V}-glutathione complex,[2,36,40-43] formed by reduction of TcO_4^- with glutathione, was tested with the RP-IPC system described in section 7.2.7. A sample was analysed 4 h after the start of the reaction. The chromatogram showed a peak with an elution volume of 3.3 mL (k' = 1.6), which could be attributed to Tc^{V}-glutathione. However, no corresponding TcX peak was present in the chromatogram (Figure 7.3B) at this elution volume, thus suggesting the absence of Tc-glutathione in spinach plants.

The structure of phytochelatins,[39] a set of heavy metal complexing peptides, was characterised as (γ-L-glutamyl-L-cysteinyl)$_n$-glycine (n = 2-11) [(γ-L-Glu-L-Cys)$_n$-Gly]. These peptides are induced in plants by some metals of the transition and main groups by activation of the enzyme PC synthase. Up to now, no Tc-phytochelatine species have been found, but metallothionein, a related heavy metal complexing species in animals, binds Tc^{V} in a thermodynamically stable and kinetically inert thiolate complex inert to oxidation and transchelation in vivo.[44-47]

In recent years considerable effort has been focused on binding ^{99m}Tc to proteins for diagnostic imaging of human diseases. Cysteinyl groups seem the major binding site for $[Tc^{V}O]^{3+}$ in peptides[48] and proteins.[49,50] Binding of ^{99m}Tc to amino or amide nitrogen atoms[51] or to imidazole moieties of histidine[50] has been observed, but only when cysteine groups are not available. The bond between reduced Tc and protein-thiolate is very strong and not very vulnerable to transchelation or oxidation.[49]

7.4 CONCLUSION

Characterisation of Tc species in spinach plants has been partly realised using a combination of selective extraction and chromatographic techniques. Since extractions deliver low-resolution data, only a crude separation of Tc species in classes could be obtained. However, combination with high-resolution chromatographic systems (SEC and RP-IPC), yielded more information about individual Tc species. The suggestion,[2,9-11] that Tc^{V}-cysteine is one of the major Tc species in spinach leaf homogenate, could be proven. Approximately 25% of total Tc could be attributed to Tc^{V}-cysteine and about 7% to TcO_4^-, the Tc species originally present in the nutrient solution.

It is obvious that standards are necessary for identification purposes in the approach chosen. Direct structural information with techniques such as mass spectrometry,[52] EXAFS,[50] ^{99}Tc-NMR,[53,54] and EPR[54,55] would be preferable, but since these techniques still have technical limitations and/or sensitivity problems, at present, the approach chosen seems the way forward in elucidation of TcX species. Near future research for the identification of Tc species separated has to focus on the *in-vitro* synthesis of other Tc compounds which might be expected, combined with the development of hyphenated methods such as HPLC/MS.

7.5 REFERENCES

(1) Lembrechts, J. F.; Desmet, G. M.; Overbeek, H. *Environ. Exp. Bot.* **1985**, *25*, 355-360.
(2) Lembrechts, J. F.; Desmet, G. M. *Health Phys.* **1989**, *57*, 255-262.
(3) Van Loon, L. R.; Desmet, G. M.; Cremers, A. *Health Phys.* **1989**, *57*, 309-314.
(4) Cataldo, D. A.; Garland, T. R.; Wildung, R. E. in *Technetium in the*

environment; Desmet, G. M., Myttenaere, C., Eds.; Elsevier: London, 1986; pp 265-280.
(5) Garten, Jr., C. T.; Lomax, R. D. *Health Phys.* **1989**, *57*, 299-307.
(6) Henrot, J. *Health Phys.* **1989**, *57*, 239-245.
(7) Harms, A. V.; Van Elteren, J. T.; Claessens, H. A. *J. Chromatogr. A* **1996**, *755*, 219-225.
(8) Harms, A. V.; Van Elteren, J. T. *J. Radioanal. Nucl. Chem.* **1998**, *227*, in press.
(9) Roucoux, P. *Etude du comportement du technétium (polluant radioactif) chez le soja en milieu contaminé ou non en cadmium (polluant conventionnel)*, Ph.D. thesis; Louvain-la-Neuve, 1980.
(10) Delmotte, A. *Comportement écophysiologique d'Anabaena cylindrica Lemn (cyanobactérie en présence de technétium-99 (polluant nucléaire)*, Ph.D. thesis; Louvain-la-Neuve, 1981.
(11) Myttenaere, C.; Vandecasteele, C. M.; Roucoux, P.; Lietart, A.; Mousny, J. M. in *Technetium in the environment*; Desmet, G. M., Myttenaere, C., Eds.; Elsevier: London, 1986; pp 281-293.
(12) Johannsen, B.; Syhre, R.; Spies, H.; Münze, R. *J. Nucl. Med.* **1978**, *19*, 816-824.
(13) Spies, H.; Johannsen, B.; Münze, R.; Unverferth, K. *Z. Anorg. Allg. Chem.* **1978**, *447*, 215-220.
(14) Hnatowich, D. J.; Virzi, F.; Fogarasi, M.; Rusckowski, M.; Winnard, Jr., P. *Nucl. Med. Biol.* **1994**, *21*, 1035-1044.
(15) Wang, A.-Y.; Lin, J.-L.; Lo, J.-G.; Tsai, Z.-T. *J. Radioanal. Nucl. Chem.* **1992**, *159*, 129-144.
(16) Wang, A.-Y.; Liu, R.-S. *Nucl. Med. Biol.* **1995**, *22*, 937-942.
(17) Farago, M. E. in *Plants and the chemical elements*; Farago, M. E., Ed.; VCH: Weinheim, 1994; pp. 265-283.
(18) Bowen, H. J. M.; Cawse, P. A.; Thick, J. *J. Exp. Bot.* **1962**, *13*, 257-267.
(19) Robinson, T. *The organic constituents of higher plants*, 4th Ed.; Cordus Press: North Amherst, 1980.
(20) Schümichen, C.; Koch, K.; Kraus, A.; Kuhlicke, G.; Weiler, K. *J. Nucl. Med.* **1980**, *21*, 1080.
(21) Hwang, L. L.-Y.; Ronca, N.; Solomon, N. A.; Steigman, J. *Int. J. Appl. Radiat. Isot.* **1984**, *35*, 825-830.

(22) Zhou, T.; Hirth, W. W.; Heineman, W. R.; Deutsch, E. *Nucl. Med. Biol.* **1988**, *15*, 493-498.
(23) Krijger, G. C.; Van Vliet, P. M.; Wolterbeek, H. T. *J. Exp. Bot.* **1998**, in preparation.
(24) Davison, A.; Trop, H. S.; DePamphilis, B.; Jones, A. G. *Inorg. Synth.* **1982**, *21*, 160-162.
(25) Johannsen, B.; Spies, H. in *Topics in Current Chemistry 176: Technetium and Rhenium*; Yoshihara, K., Omori, T., Eds.; Springer: Berlin, 1996; pp 77-121.
(26) Vandecasteele, C. M.; Garten, Jr., C. T.; Van Bruwaene, R.; Janssens, J.; Kirchmann, R.; Myttenaere, C. in *Speciation of fission and activation products in the environment*; Bulman, R. A.; Cooper, J. R., Eds.; Elsevier: London, 1986; pp 368-380.
(27) Garten, Jr., C. T.; Myttenaere, C.; Vandecasteele, C. M.; Kirchmann, R.; Van Bruwaene, R. in *Technetium in the environment* Desmet, G. M., Myttenaere, C., Eds.; Elsevier: London, 1986; pp 319-322.
(28) Myttenaere, C.; Daoust, C.; Roucoux, P. *Environ. Exp. Bot.* **1980**, *20*, 415-419.
(29) Kotegov, K. V.; Pavlov, O. N.; Shvedov, V. P. in *Advances in inorganic chemistry and radiochemistry*, Vol. 11; Emeléus, H. J.; Sharpe, A. G., Eds.; Academic press: London, 1968; pp 1-90.
(30) Franklin, K. J.; Howard-Lock, H. E.; Lock, C. J. L. *Inorg. Chem.* **1982**, *21*, 1941-1946.
(31) Yokoyama, A.; Horiuchi, K. *Int. J. Appl. Radiat. Isot.* **1982**, *33*, 929-936.
(32) Marzilli, L. G.; Banaszczyk, M. G.; Hansen, L.; Kuklenyik, Z.; Cini, R.; Taylor, Jr., A. *Inorg. Chem.* **1994**, *33*, 4850.
(33) Cataldo, D. A.; Garland, T. R.; Wildung, R. E.; Fellows, R. J. *Health Phys.* **1989**, *57*, 281-287.
(34) Schwenn, J. D. *Z. Naturforsch.* **1994**, *49c*, 531-539.
(35) Leustek, Th. *Physiol. Plant.* **1996**, *97*, 411-419.
(36) Lembrechts, J. F. *Ph. D. thesis*; U. I. Antwerpen, 1986.
(37) *Coenzymes and cofactors, Vol.III: Glutathione, Chemical, Biochemical, and Medical Aspects, part A*; Dolphin, D.; Poulson, R., Avramovic, O., Eds.; Wiley: New York, 1989.
(38) *Glutathione: Metabolism and Physiological functions*; Viña, J., Ed.;

CRC Press: Boca Raton, 1990.
(39) Zenk, M. H. Gene **1996**, *179*, 21-30.
(40) Hnatowich, D. J.; Mardirossian, G.; Rusckowski, M.; Fogarasi, M.; Virzi, F.; Winnard, Jr., P. *J. Nucl. Med.* **1993**, *34*, 109-119.
(41) Fritzberg, A. R.; Lyster, D. M.; Dolphin, D. H. *Int. J. Nucl. Med. Biol.* **1978**, *5*, 87-92.
(42) Fritzberg, A. R.; Lyster, D. M.; Dolphin, D. H. *J. Labelled Compd. Radiopharm.* **1979**, *16*, 190-191.
(43) Ercan, M. T.; Bekdik, C. F.; Sarizi, T. *Int. J. Appl. Radiat. Isot.* **1978**, *29*, 697-698.
(44) Morelock, M. M.; Cormier, T. A.; Tolman G. L. *Inorg. Chem.* **1988**, *27*, 3137-3140.
(45) Jones, W. B.; Elgren, T. E.; Morelock, M. M.; Elder, R. C.; Wilcox, D. E. *Inorg. Chem.* **1994**, *33*, 5571-5578.
(46) Goudard, F.; Durand, J.-P.; Galey, J.; Piéri, J.; Masson, M.; George, S. *Mar. Biol.* **1991**, *108*, 411-417.
(47) Brown, B. A.; Drozynski, C. A.; Dearborn, C. B.; Hadjian, R. A.; Liberator, F. A.; Tulip, T. H.; Tolman, G. L.; Haber, S. B. *Anal. Biochem.* **1988**, *172*, 22-28.
(48) Johannsen, B.; Jankowsky, R.; Noll, B.; Spies, H.; Reich, T.; Nitsche, H.; Dinkelborg, L. M.; Hilger, C. S.; Semmler, W. *Appl. Radiat. Isot.* **1997**, *48*, 1045-1050.
(49) Rhodes, B. A. *Nucl. Med. Biol.* **1991**, *18*, 667-676.
(50) Zamora, P. O.; Rhodes, B. A. *Bioconjugate Chem.* **1992**, *3*, 493-498.
(51) Vanbilloen, H. P.; Bormans, G. M.; De Roo, M. J.; Verbruggen, A. M. *Nucl. Med. Biol.* **1995**, *22*, 325-338.
(52) Costello, C. E.; Brodack, J. W.; Jones, A. G.; Davison, A.; Lee Johnson, D.; Kasina, S.; Fritzberg, A. R. *J. Nucl. Med.*, **1983**, *24*, 353-355.
(53) O'Connell, L. A.; Pearlstein, R. M.; Davison, A.; Thornback, J. R.; Kronauge, J. F.; Jones, A. G. *Inorg. Chim. Acta* **1989**, *161*, 39-43.
(54) Van Vliet, P. M.; Krijger, G. C. unpublished results.
(55) Abram, U.; Kirmse, R.*Radiochim. Acta* **1993**, *63*, 139-143.

Chapter 8

Conclusions

8.1 INTRODUCTION

In this chapter, main conclusions are presented on the analytical aspects of technetium speciation in spinach plants. As mentioned in Chapter 1, elemental speciation is often troublesome, because of the problems associated with measurements of unknown species, generally present in low concentrations, which certainly applies to “natural” levels of technetium. Moreover, reactions of species during the determination procedure will result in artefacts and will affect the final results. Elemental speciation in plant material generally requires some form of pretreatment before the actual isolation, separation, and determination of species. The speciation procedures described in this thesis consist of a plant material homogenisation step and chromatographic analysis of the filtrate or supernatant of this homogenate. As the homogenisation and separation steps may result in decomposition of existing species and formation of new species, results of speciation studies should be checked on artefacts. Artefacts cannot always be excluded, but certain precautions can be made.

8.2 SPECIATION METHOD DEVELOPMENT

Firstly, one should critically evaluate the composition of the homogenisation buffer. Inattentive addition of certain protective compounds such as EDTA, thiols and buffers with metal complexing abilities (e.g., Tris) may result in the *in vitro* formation of new species, not present *in vivo*. The work described in Chapter 3 illustrates the unforeseen role "protective" compounds may play in homogenisation buffers. Three technetium species, i.e., TcO_4^-, Tc-cyclam, and Tc-DTPA, were incubated with 2-mercaptoethanol, a thiol which is used in plant protein isolation procedures as an inhibitor of polyphenol oxidase. However, this 2-mercaptoethanol and other thiols may act as a ligand for reduced technetium or as a reducing agent for pertechnetate, a weak oxidising agent, itself. The results of this study indicate that Tc-DTPA, which is widely used as a radiopharmaceutical, in fact reacts with 2-mercaptoethanol. No reaction with TcO_4^- and the more inert Tc-cyclam was observed.

Secondly, one should select "mild" separation conditions in order to avoid transchelation, dissociation, and irreversible adsorption during the separation of the metal species to be isolated. Therefore, in Chapter 4 attention is focused on the development of such a separation technique for Tc species. The same Tc species as in Chapter 3, i.e., TcO_4^-, Tc-cyclam, and Tc-DTPA, were used in this chapter to illustrate the separation potential of size-exclusion chromatography (SEC). Several non-size effects, i.e., interactions of the solute with the stationary phase, which affect the relation between the elution volume and the size of a solute, were observed: ion-exclusion, hydrophobic interaction, cation exchange, and adsorption. These effects will obscure molecular weight determinations, which are based on elution volume alone. It was found that hydrophobic interaction of TcO_4^- with the stationary phase was useful in separating TcO_4^- from both Tc species. This non-size effect was found to increase with NaCl concentration in the mobile phase and to decrease with higher column temperatures. This non-size effect was further exploited in Chapter 5 for the separation of TcO_4^- and TcX (plant-induced Tc species).

Thirdly, one should have a method to check the integrity of species during the speciation procedure. In Chapter 5 a dual radiotracer technique for probing of speciation artefacts such as reduction, oxidation, and selective procedural loss of species, was developed for a two species system. This novel technique permits the determination of the *in situ* leaf concentration of two technetium species, i.e., $^{99}TcO_4^-$ and ^{99}TcX as a function of incubation time with $^{99}TcO_4^-$ in the substrate. Addition of $^{95m}TcO_4^-$ or $^{99m}TcO_4^-$, both γ-emitting species, to the substrate allowed easy radiotracing, while addition of $^{99m}TcO_4^-$ or ^{95m}TcX during the analytical homogenization procedure enabled us to correct for procedural losses and redox conversions. Major findings in this chapter are significant procedural losses of $^{99}TcO_4^-$, increasing with incubation time, and insignificant redox conversions. In the experimental set-up used, a few assumptions have been made: (i) there is an instant equilibrium between the tracers and the tracee both in the nutrient solution and in the homogenate; (ii) the ratio of sub-species of TcX is independent of incubation time; (iii) the ratio of the sub-species is independent of the Tc concentration in the nutrient solution. The dual tracer technique developed in this chapter may be extended to speciation of other elements, having at least two suitable radioisotopes (e.g., Mn, Co, As, Se, Sn, Sb, and I). Moreover, this technique may also be used with enriched stable isotopes combined with mass spectrometry.

8.3 DIFFERENTIATION OF TcX SPECIES

In the last part of this thesis, i.e., Chapters 6 and 7, attention is focused on differentiation of TcX species. As the "mild" separation method SEC used to separate TcO_4^- and TcX in Chapter 5 has not enough resolving power to separate individual TcX species, it was necessary to seek other types of chromatography, e.g., reversed-phase ion-pair chromatography (RP-IPC). The chromatographic behaviour of TcO_4^-, Tc-cyclam, and Tc-DTPA under RP-IPC separation conditions (using tetrabutylammonium acetate as ion-pair reagent and methanol as organic modifier at pH 5.9) was tested. The retention behaviour of TcO_4^- strongly depends on the concentration of ion-pair reagent or organic modifier. However, adverse interactions of cationic Tc-cyclam and anionic Tc-DTPA with the stationary phase, resulting in poor peak shapes and low recoveries, were also observed. The reason for this behaviour may be found in dissociation and/or binding of the species to residual silanol groups. The addition of blocking agents, such as EDTA or other amines, which are normally used to reduce these problematic interactions, is, for obvious reasons, not feasible for this type of separation. One of the TcX sub-species, which was adsorbed on the stationary phase of the RP-IPC column, was excluded on a SEC column, due to charge or, more likely, size. This species may be attributed to Tc bound to proteins with a molecular mass exceeding 5 kDalton.

In Chapter 7, the TcX speciation study is extended to a combination of selective extraction techniques and RP-IPC or SEC. The results in this chapter indicate that the RP-IPC separation (using trifluoroacetic acid and methanol at pH 2.2), which is a common separation technique for proteins and peptides, is also suitable for separating inert Tc species. The combination of extraction and chromatographic techniques revealed several classes of TcX in spinach plants. Approximately 17% of the Tc present in spinach leaves was pelleted after centrifugation of the homogenate. This fraction may be attributed to Tc bound to undissolved cell-wall polysaccharides. The "yellow" Tc-cysteine, probably $[Tc^{V}O(cys)_2]$, which was synthesised *in vitro*, was identified with both chromatographic techniques as one of the major sub-species of TcX. Approximately 25% of total Tc could be attributed to this Tc species. The Tc-glutathione complex, which was also synthesised in vitro, could not be detected in the supernatant of homogenised spinach leaves. The other TcX sub-species discovered may consist of Tc bound to proteins, Tc bound to peptides such as phytochelatine, or Tc bound to small ligands such as organic or amino acids other than cysteine, e.g., Tc-nicotianamine, Tc-oxalate, Tc-citrate, or Tc-histidine. The extraction data combined with the results in Chapter 6 indicate that approximately 26% of total Tc consists of Tc bound to proteins. Approximately 25% of total Tc consists of highly polar non-protein Tc species, e.g. Tc bound to peptides or small ligands. Finally, about 7% of the total Tc could be attributed to TcO_4^-, the Tc species originally present in the nutrient solution.

The weak point of the approach chosen remains the identification of compounds which so far is by matching with synthesised Tc compounds. However, real technical advances in speciation methods are not to be expected in the near future so it seems that the way forward in Tc speciation in plants is by trial-and-error identification with synthesised Tc compounds. Only when hyphenated systems like HPLC-MS will become sensitive enough a breakthrough in technetium speciation might take place.

8.4 METABOLIC BEHAVIOUR OF TcO_4^-

The metabolic behaviour of TcO_4^- and reduced Tc species in plants remains to be elucidated. However, the progress in related work concerning the metabolic behaviour of ^{99m}Tc-radiopharmaceuticals in animal and human systems is encouraging. It seems reasonable to assume that in plants TcO_4^- is taken up, transported to the leaves, and subsequently reduced. These reduced Tc-cores, e.g. $[Tc^{V}O]^{3+}$, will bind to (in)organic ligands which are present in ubiquitous forms in plant cells. It is clear that the multitude of ligands might result in a large variety of Tc species.

Some results presented in this thesis might shed a light on the metabolic pathway of TcO_4^- to TcX. The existence of Tc^{V}-cysteine as one of the major sub-species of TcX, despite the very low concentration of free cysteine in spinach, suggests a specific interaction of reduced Tc with cysteine. This might be an indication for a relation between the metabolism of sulphur and technetium. TcO_4^- might be taken up as a SO_4^{2-}-analogue, being reduced in chloroplasts to Tc^{V} and subsequently bound to L-cysteine, the primary organic metabolite of SO_4^{2-} and precursor of all sulphur-containing organic plant compounds, including glutathione. The binding of Tc to glutathione, the "storage" for excess sulfur, which has been reported for metals such as Zn, Hg, Pb, Ni, and Cd was not observed in this work.

It is interesting that the reported toxic effects of Tc in plants, e.g., lipid peroxidation (oxidative stress), resulting in membrane damage, are identical to effects caused by "soft" metal ions such Cu^{II}, Hg^{II}, Pb^{II}, and Cd^{II}. Despite the apparent difference in uptake and initial metabolism, Tc in toxic amounts might interfere with the glutathione metabolism (by binding to its precursor), and therefore interfere with oxidative and/or enzymatic processes in the cell.

8.5 REFLECTION

Summarising we can say that onsets have been made to a suitable controllable procedure for speciation of technetium in spinach leaves in view of: i) the composition of the homogenisation buffer and mobile phase; ii) monitoring of potential artefacts during the pretreatment procedure; and iii) "mild" chromatographic separation and selective extraction techniques for enhanced resolution. This resulted in separation of several technetium species in the filterable fraction of the leaf homogenate. TcO_4^- and Tc^V-cysteine could be identified unequivocally, but other compounds still need to be elucidated with the trial-and-error method by matches with synthesised compounds. However, the unfilterable fraction remains to be studied but this is a new field of research (solid matter speciation) to be tackled.

It should be stressed that in this study technetium levels were substantially higher than encountered in "natural" samples. However, since the levels used in this study are sub-toxic, there is no ground to assume that "natural" levels will behave differently. Studying "natural" levels with the proposed technique is possible, however, it requires substantially more effort to detect such low "natural" levels of ^{99}Tc.

The speciation methods developed in this thesis are currently in use in a multidisciplinary study which involves plant-physiological aspects of technetium in tomato plants and uptake and transformation kinetics of technetium in duckweed. Moreover, the approach followed is not specific for technetium in plants, but applies to most speciation studies in life and environmental sciences.

Summary

Technetium (τεχνητος; Greek artificial; Tc) is a 4*d*-block metal with only radioactive isotopes. It was discovered in 1937 in Italy as the first man-made element. The radionuclide ^{99}Tc is produced in appreciable amounts by fission of ^{235}U and ^{239}Pu in the nuclear fuel cycle. The large fission yield (6.1%), the long half-life (213,000 y) combined with the considerable mobility in environmental systems of pertechnetate (TcO_4^-) and the remarkable ability of technetium to concentrate in marine and freshwater wildlife and terrestrial plants, has stimulated the interest in the radioecology of this element. The history, nuclear properties, chemistry, the role in nuclear medicine, speciation, environmental and analytical aspects of technetium are briefly presented in Chapter 2.

To understand the behaviour of technetium in biological and environmental systems, it is necessary to obtain information on the chemical forms (species) of technetium involved. Speciation, i.e., determination of metal species in plants is difficult for several reasons: i) species are unknown; ii) species may be altered as a result of the speciation procedure; and iii) the extremely low "natural" level of technetium in the plant. Limitations of present procedures lie in the integrity of sample pretreatment methods and in the availability of sufficiently assessed separation methods.

Metal speciation in plant material generally requires some form of pretreatment before the actual isolation, separation, and determination of species. Reduction, oxidation, hydrolysis, transchelation induced by the pretreatment procedures may cause profound changes to certain species. The multitude of organic and inorganic constituents in plants makes it difficult to predict exactly what problems might be encountered.

Depending on the nature of the compounds to be isolated, various protective compounds may be added to the homogenisation medium. However, inattentive buffers, osmotica, metal ion chelators, detergents, reducing agents, may result in the *in vitro* formation of new species, not present *in vivo*. The work described in Chapter 3 illustrates the unforeseen role "protective" compounds might play in homogenisation buffers. Three technetium species, i.e., TcO_4^-, Tc-cyclam, and Tc-DTPA, were incubated with 2-mercaptoethanol (2-ME), a thiol which is used in plant protein isolation procedures as an inhibitor of polyphenol oxidase. However, 2-ME as well as other thiols may act as a ligand for reduced technetium or as a reducing agent for pertechnetate, a weak oxidising agent, itself. It is shown that Tc-DTPA, which is widely used as a radiopharmaceutical, does in fact react with 2-ME, presumably resulting in the formation of a $[Tc^{V}O\text{-}(2\text{-}ME)_2]$ complex. No reaction with TcO_4^- and the more inert Tc-cyclam was observed.

To reduce adverse chromatographic behaviour resulting from transchelation, dissociation, and irreversible adsorption of labile Tc species, "mild" separation conditions and the use of an inert chromatographic system are required. The same Tc species as in Chapter 3, i.e., TcO_4^-, Tc-cyclam, and Tc-DTPA, were used in this chapter to illustrate the separation potential of size-exclusion chromatography (SEC). Several non-size effects, i.e., interactions of the solute with the stationary phase, which affect the relation between the elution volume and the size of a solute, were observed: ion-exclusion, hydrophobic interaction, cation exchange, and adsorption. The extent of these non-size effects has a close relation with the nature of the Tc species, the temperature, and the nature of the mobile and stationary phase. One of the stationary phases, i.e., HEMA, was used in Chapter 5 to separate TcO_4^- from plant-induced Tc species (TcX).

Chapter 5 is dedicated to the development of an analytical procedure for Tc speciation in spinach plants using a dual Tc radiotracer technique. This technique, which was developed for a two species system, provides corrections for procedural losses and redox conversions of both TcO_4^- and TcX. To this end, $^{99m}TcO_4^-$ or ^{95m}TcX were added during the analytical homogenization procedure of plants grown on nutrient solutions containing $^{99}TcO_4^-$, which was traced with $^{95m}TcO_4^-$ or $^{99m}TcO_4^-$, respectively. This technique permits the determination of the *in situ* leaf concentration of two technetium species, i.e., $^{99}TcO_4^-$ and ^{99}TcX as a function of incubation time with $^{99}TcO_4^-$ in the substrate. Major findings in this chapter are significant procedural losses of $^{99}TcO_4^-$, increasing with incubation time, and insignificant redox conversions.

$^{99}TcO_4^-$, increasing with incubation time, and insignificant redox conversions. The dual tracer technique developed in this chapter may be extended to speciation of other elements, having at least two suitable radioisotopes (e.g., Mn, Co, As, Se, Sn, Sb, and I).

Chapters 6 and 7 are dedicated to a further investigation of the speciation of TcX itself with extraction and chromatographic techniques. As the "mild" separation method SEC used to separate TcO_4^- and TcX in Chapter 5 has not enough resolving power to separate individual TcX species, it was necessary to seek other types of chromatography, e.g., reversed-phase ion-pair chromatography (RP-IPC). In Chapter 6 the chromatographic behaviour of TcO_4^-, Tc-cyclam, and Tc-DTPA under RP-IPC separation conditions was tested. The retention behaviour of TcO_4^- strongly depends on the concentration of ion-pair reagent or organic modifier. However, adverse interactions of cationic Tc-cyclam and anionic Tc-DTPA with the stationary phase, resulting in poor peak shapes and low recoveries, were also observed. The reason for this behaviour may be found in dissociation and/or binding of the species to residual silanol groups. One of the TcX sub-species, which was adsorbed on the stationary phase of the RP-IPC column, was excluded on a SEC column, due to charge or, more likely, size. This species may be attributed to Tc bound to proteins with a molecular mass exceeding 5 kDalton.

In Chapter 7, the TcX speciation study is extended to a combination of selective extraction techniques and RP-IPC or SEC. The combination of extraction and chromatographic techniques reveals several classes of TcX in spinach plants. Approximately 17% of the Tc present in spinach leaves may be attributed to Tc bound to insoluble cell-wall polysaccharides. By comparison of the chromatographic behaviour of TcX with *in vitro* synthesised Tc complexes, it is shown that Tc^V-cysteine, probably $[Tc^VO(cys)_2]$, is a major sub-species of TcX. Approximately 25% of total Tc could be attributed to this Tc species. The extraction data combined with the results in Chapter 6 indicate that approximately 26% of total Tc consists of Tc bound to proteins. Approximately 25% of total Tc consists of Tc bound to peptides such as phytochelatine, or Tc bound to small ligands such as organic or amino acids other than cysteine. Finally, about 7% of the total Tc could be attributed to TcO_4^-, the Tc species originally present in the nutrient solution. The binding of Tc to glutathione, the "storage" for excess sulphur, which might be expected, was not observed in this despite the very low concentration of free cysteine in spinach, suggests a relation between the metabolism of sulphur and technetium. However, the complete metabolic behaviour of TcO_4^- and species of reduced Tc in plants remains to be elucidated.

Chapter 8 evaluates the results obtained in view of the development of an analytical procedure for speciation of technetium in spinach leaves. Factors to be taken into account for a suitable controlled procedure are: i) composition of the homogenisation buffer, stationary and mobile phase; ii) "mild" chromatographic separation conditions and selective extraction conditions; and iii) monitoring of potential artefacts during the pretreatment procedure. The techniques developed in this thesis may be extended to speciation of other elements in other matrices.

Samenvatting

Technetium (τεχνητος; Grieks voor kunstmatig; afkorting Tc), een element waarvan alle isotopen radioactief zijn, komt vrijwel niet in de natuur voor. Het werd in 1937 ontdekt als het eerste kunstmatig gemaakte element. Het is een (4*d*-overgangs)metaal, waarvan het bestaan al eerder op basis van Periodiek Systeem was voorspeld. In de jaren veertig bleek dat het isotoop van technetium met het massagetal 99 (^{99}Tc) in relatief grote hoeveelheden wordt gevormd na kernsplijting van uranium-235 en plutonium-239. Het radionuclide ^{99}Tc, dat vrijwel alleen β^--straling uitzendt, is gezien de hoge splijtingsopbrengst (6,1%) gecombineerd met de lange halveringstijd (213.000 jaar) op lange termijn één van de belangrijkste splijtingsproducten in kernafval. De chemische vorm van technetium onder aërobe omstandigheden is het pertechnetaat anion (TcO_4^-), waarin technetium zich in de oxidatietoestand VII bevindt. Pertechnetaat zal vanwege de negatieve lading nauwelijks gebonden worden door verbindingen in de bodem en is daardoor mobiel in een aëroob milieu. In het inleidende hoofdstuk 2 worden een aantal aspecten van technetium behandeld: de ontdekking, de nucleaire eigenschappen, de chemie, het gebruik in de nucleaire geneeskunde, het voorkomen als splijtingsproduct, de opname door planten en bepalings- en speciatiemethoden.

Een van de eigenschappen van technetium is de hoge opname ervan door planten, hetgeen opmerkelijk is, omdat gedurende hun evolutie planten geen contact met technetium gehad kunnen hebben. Het is echter waarschijnlijk dat planten TcO_4^- aanzien voor een chemisch gelijkend nutriënt, zoals bijvoorbeeld de oxi-anionen NO_3^-, SO_4^{2-}, of MoO_4^{2-}. Door de accumulatie-eigenschap kunnen planten gebruikt worden als biomonitor voor Tc dat mogelijk uit kernafval in het milieu terecht zou kunnen komen. Na opname door de wortels wordt het technetium getransporteerd naar de bladeren en verandert de chemische vorm van het technetium door reductie en binding aan andere verbindingen (liganden). Na reductie van het pertechnetaat zullen er een aantal chemische vormen van technetium ontstaan die in de literatuur en in dit proefschrift TcX worden genoemd.

Het is onmogelijk vooraf te voorspellen welke chemische vormen van technetium (Tc-species) in een plant zullen ontstaan, omdat er zeer veel verschillende verbindingen in de plantencel (ondermeer eiwitten, peptiden, aminozuren, DNA, suikers, organische zuren en anorganische verbindingen) technetium kunnen binden. Kennis van de chemische vormen van het aanwezige Tc, ook wel genoemd de speciatie van Tc, zou iets kunnen zeggen over de metabolisatie van Tc in de plant.

Voordat met de scheiding en karakterisering van de Tc-species begonnen kan worden, moeten deze eerst worden geïsoleerd uit het plantmateriaal. Daartoe wordt het plantmateriaal, in dit onderzoek spinaziebladeren, in een oplossing door vermaling gehomogeniseerd, gevolgd door filtratie of centrifugatie van de ontstane suspensie, waarna de resterende oplossing wordt geanalyseerd met behulp van vloeistofchromatografie. Echter, doordat homogenisatie van het plantmateriaal verbindingen uit verschillende cel-compartimenten met elkaar in contact brengt, zullen reacties van Tc-species tijdens de isolatie- en analyseprocedure onjuiste informatie over de aanwezigheid van in de plant gevormde Tc-species opleveren. In dit proefschrift worden een aantal voorzorgsmaatregelen behandeld om een bijdrage aan de oplossing van het bovenstaande probleem te leveren.

Ten eerste wordt er gekeken naar de samenstelling van de homogenisatie-buffer, waarin verscheidene verbindingen ter bescherming van de te isoleren species kunnen worden toegevoegd. In hoofdstuk 3 wordt een dergelijke verbinding, 2-mercaptoethanol (2-ME), die aan homogenisatie- en isolatie-buffers wordt toegevoegd om verbindingen te beschermen tegen oxidatieve reacties, getest op drie Tc-modelverbindingen. Het blijkt dat één van de modelverbindingen, namelijk het Tc-DTPA complex, met 2-ME reageert tot een nieuwe verbinding. 2-ME blijkt dus niet alleen te beschermen tegen oxidatie, maar blijkt ook te reageren met Tc-verbindingen.

Ten tweede wordt er een zo mild mogelijke scheidingsmethode voor de Tc-species ontwikkeld. Het scheidend vermogen van de gebruikte methode, vloeistofchromatografie, is gebaseerd op de verschillen in interacties van de te scheiden verbindingen met de mobiele en de stationaire fase gedurende de periode dat zij opgelost in de mobiele fase door een met een poreuze stationaire fase gevulde kolom stromen. Door de temperatuur, de mobiele en/of de stationaire fase te variëren kan de verblijftijd van elk van de verbindingen op de kolom worden beïnvloed, waardoor mogelijk een scheiding kan worden veroorzaakt. Een succesvolle scheiding van mogelijk labiele verbindingen vereist een zo mild mogelijke scheidingsmethode. Daarom werd gebruik gemaakt van een milde techiek, die normaliter alleen op basis van molecuulgrootte scheidt (size-exclusion chromatografie). Kleinere moleculen kunnen dieper in de poriën van een inerte stationaire fase doordringen dan grotere moleculen en hebben zodoende een groter volume op de kolom ter beschikking, hetgeen leidt tot een grotere verblijftijd op de kolom. Deze manier van scheiden, die veel wordt gebruikt om eiwitten en polymeren te scheiden en vervolgens op molecuulgrootte te karakteriseren, kan echter worden beïnvloed door zogenaamde "non-size" effecten. Deze effecten die zorgen voor een kortere dan wel langere verblijftijd op de kolom dan op basis van grootte verwacht mag worden zijn afhankelijk van de temperatuur, de aard van de te scheiden verbinding, de mobiele en/of stationaire fase. In hoofdstuk 4 wordt door variatie van de stationaire fase, de temperatuur en de zoutsterkte van de mobiele fase een aantal verschillende "non-size" effecten bestudeerd. Het blijkt dat een aantal Tc modelverbindingen die weinig verschillen in molecuulgrootte (maar wel in lading en hydrofobiciteit) door gebruikmaking van deze "non-size" effecten te scheiden zijn.

In hoofdstuk 5 wordt aandacht besteed aan controlemethoden die het mogelijk maken te bekijken of de verkregen speciatieresultaten ook inderdaad representatief zijn voor de *oorspronkelijke* speciesverdeling. In dit hoofdstuk wordt $^{99}TcO_4^-$ aan de voedingsoplossing van spinazieplanten toegevoegd. Om de detectie van het ^{99}Tc te vergemakkelijken, wordt er een ander Tc-radioisotoop dat γ-straling uitzendt (^{99m}Tc met een halfwaardetijd van 6.02 uur of ^{95m}Tc met een halfwaardetijd van 61 dagen) als tracer aan de voedingsoplossing toegevoegd. Zoals bij alle tracertoepassingen wordt er vanuit gegaan dat het chemisch gedrag van de verschillende radioisotopen, ^{95m}Tc, ^{99m}Tc en ^{99}Tc, identiek is. Met de in hoofdstuk 4 ontwikkelde chromatografische techniek is het mogelijk om TcO_4^-, de chemische vorm waarin het technetium aan de spinazieplanten wordt toegevoegd, te scheiden van TcX, de in de spinaziebladeren gevormde Tc-species. Om deze analyse echter mogelijk te maken moeten de Tc-species (TcO_4^- en TcX) uit de bladeren ontsloten worden. Door nu TcO_4^- of TcX in de vorm van een derde Tc-isotoop aan de ontsluitingsbuffer toe te voegen kan gecontroleerd worden of *tijdens* de ontsluitingsprocedure species worden omgezet of dat er selectieve verliezen van een bepaald species optreden. In de gekozen opzet is aan planten die met $^{95m}TcO_4^-$ en $^{99}TcO_4^-$ zijn opgegroeid, ter controle van het TcO_4^- tijdens de ontsluitingsprocedure $^{99m}TcO_4^-$ toegevoegd, terwijl aan planten die met $^{99m}TcO_4^-$ en $^{99}TcO_4^-$ zijn opgegroeid, ^{95m}TcX ter controle van TcX is toegevoegd. Deze ^{95m}TcX-tracer was in een eerder experiment geïsoleerd uit planten die waren opgegroeid met $^{95m}TcO_4^-$. Uit de experimenten blijkt dat TcO_4^- in de plant wordt omgezet tot TcX en dat er tijdens de gekozen ontsluitingsprocedure een verwaarloosbare hoeveelheid TcO_4^- wordt omgezet in TcX. Tevens blijkt dat er een relatief groot gedeelte van het TcO_4^- adsorbeert aan de filters die worden gebruikt om het bladerenhomogenaat geschikt te maken voor injectie in de chromatograaf. Dit betekent dat zonder correctie er een relatief te lage hoeveelheid van het TcO_4^- in het plantmateriaal zou zijn bepaald.

In de hoofdstukken 6 en 7 wordt geprobeerd te bepalen uit welke chemische vormen het TcX bestaat om zodoende aanwijzingen te verkrijgen over het metabolisme van technetium in planten. Daartoe worden selectieve extractie en twee soorten vloeistofchromatografie toegepast, namelijk reversed-phase ion-pair chromatografie (RP-IPC) en de al beschreven size-exclusion chromatografie (SEC). RP-IPC is een minder milde scheidingstechniek dan SEC, maar heeft op haar beurt een groter scheidend vermogen dan de laatste techniek. De stationaire fase bestaat bij RP-IPC meestal uit silica met een hydrofobe bekleding (in dit geval C_{18} ketens). Het ion-pair reagens wordt toegevoegd aan de mobiele fase, waarna, na injectie van de te scheiden verbindingen, het interactie met deze verbindingen vertoont. Deze interactie, waarin de vorming van ionparen een rol speelt, heeft een groot aandeel in het scheidend vermogen van deze techniek. Helaas zorgen de scheidings-omstandigheden, waarbij in dit geval een mengsel van een waterige oplossing en het organische oplosmiddel methanol wordt gebruikt, voor problemen bij het chromatografische gedrag van de Tc-modelverbindingen Tc-DTPA, Tc-cyclam en een gedeelte van de TcX species. Er treedt adsorptie van technetium aan de stationaire fase op en de elutiepieken zijn slecht van vorm en zeer breed. Dit wijst op reacties van de te scheiden verbindingen met het chromatografische systeem tijdens de scheiding.

In hoofdstuk 7 wordt de TcX-speciatie uitgebreid tot een combinatie van selectieve extractietechnieken met RP-IPC en/of SEC. Deze combinatie van technieken maakte een indeling in een aantal klassen van TcX mogelijk. Ongeveer 17% van de Tc-species aanwezig in de spinaziebladeren bestaat uit Tc gebonden aan onoplosbare celwand polysaccharides. Door vergelijking van het chromatografische gedrag van TcX met *in vitro* gesynthetiseerde Tc-complexen, kon aannemelijk worden gemaakt dat een Tc^{V}-cysteïne complex, waarschijnlijk $[Tc^{V}O(cys)_2]$, één van de belangrijkste subspecies van TcX is. Ongeveer 25% van het aanwezige Tc kon worden toegewezen aan dit Tc-species. Het bestaan van Tc^{V}-cysteïne als een van de belangrijkste TcX subspecies, ondanks de zeer lage concentratie van vrij cysteïne in spinazie, suggereert een relatie tussen het metabolisme van zwavel en technetium. De andere ontdekte TcX subspecies zouden kunnen bestaan uit Tc gebonden aan eiwitten (ongeveer 26%) en Tc gebonden aan peptiden (bijvoorbeeld phytochelatine) en andere kleine liganden zoals organische zuren of aminozuren anders dan cysteïne (ongeveer 25%). Tenslotte, ongeveer 7% van het aanwezige Tc kon worden toegewezen aan TcO_4^-, het Tc-species aanwezig in de voedingsoplossing van de planten. De verwachte binding van Tc aan glutathion, een peptide dat een belangrijke rol speelt in verscheidene processen in de cel, kon in dit onderzoek niet worden aangetoond.

In hoofdstuk 8 tenslotte worden de verkregen resultaten geëvalueerd. Factoren van belang voor de ontwikkeling van een geschikte controleerbare procedure betreffende de speciatie van technetium in spinaziebladeren zijn: i) samenstelling van de homogenisatiebuffer, stationaire en mobiele fase, ii) "milde" chromatografische scheiding en/of selectieve extractie en iii) monitoring van potentiële artefacten tijdens de voorbewerkingsprocedure. De verkregen resultaten hebben betrekking op technetium en spinazie, maar de beschreven aanpak kan ook gebruikt worden voor andere elementen in andere systemen.

Dankwoord

Een promotieonderzoek doe je nooit alleen. Daarom wil ik aan het eind van dit proefschrift een ieder bedanken die heeft bijgedragen aan de totstandkoming ervan. Een aantal wil ik met name noemen.

Bert Wolterbeek, die mij na de "doorstart" van mijn onderzoek in maart 1994 als "invaller" heeft begeleid, en Hans van Elteren, die vanaf zomer 1995 deze taak op dezelfde goede wijze heeft voortgezet. Mijn eerste promotor Jeroen de Goeij en tweede promotor Frans Everaerts voor hun deskundigheid op respectievelijk radiochemisch/nucleair en bio-analytisch gebied.

Mijn collega technetium-AIO's Jasper Hattink en Gerard Krijger voor de prettige samenwerking. Henk Claessens, Subramanian Iyer, Jaap de Kok, Zdenka Šlejkovec, Kristien Vercoutere en Paul van Vliet voor de interessante wetenschappelijke discussies. Patrick Chin, Koos Kroon en Ula Woroniecka voor de uitvoering van een gedeelte van het experimenteel werk. Wim den Hollander en Astrid van der Meer voor het laboratorium-, telapparatuur- en Tc-generatorbeheer. Folkert Geurink voor de hulp op computergebied. Hermien Konings en Martina Rothfusz voor de ondersteuning op secretarieel gebied. Erwin Baas, Gert Jan van den Berg, Marcel Bijvelds, Menno Blaauw, Petr Böhm, Esther Brokx, Roelof Crevecoeur, Jacco van Doornmalen, Danny van Elswijk, Marlon Fraai, Sjoerd Gelsema, Maurice Hermans, Gert-Jan Knetsch, Zvonko Kolar, Marnix ten Kortenaar, André Lagerwaard, René Leen, Ton van der Meer, Ronald Overwater, Olav Steinebach, Cees Stellema, Tona Verburg, Joost Vlek, Kees Volkers, Frank Wegh en Lennart Weltje voor de goede sfeer tijdens en soms ook na het werk.

De sectie Analytische Chemie van de KNCV en Shell Nederland wil ik bedanken voor de financiële steun, die bezoek aan een tweetal congressen mede mogelijk maakte. De T.U. Delft dank ik voor de financiële steun vanuit de materiële beleidsruimte voor de aanschaf van een "inert" HPLC-apparaat.

Tenslotte dank ik mijn ouders, zus, vrienden en familie voor hun interesse, steun en vertrouwen.

Curriculum vitae

De schrijver van dit proefschrift werd geboren op 23 september 1966 te 's-Gravenhage. In mei 1985 legde hij VWO examen aan het Haags Montessori Lyceum te 's-Gravenhage af. In september 1985 begon hij met de studie scheikunde aan de Rijksuniversiteit Leiden. In augustus 1987 behaalde hij het propaedeutisch examen scheikunde. Van november 1988 tot en met januari 1991 volgde hij een onderzoeksstage in de groep Driessen en Reedijk (bio-anorganische chemie) inzake de synthese en karakterisering van pyrazool- en (metaal)imidazool zwavelverbindingen, die als model voor koper-eiwitten zouden kunnen dienen. Van juni tot en met augustus 1990 volgde hij een onderzoeksstage bij de groep Dabrowiak (bio-anorganische chemie) aan de Syracuse University, Syracuse NY, Verenigde Staten. In augustus 1991 legde hij doctoraal examen scheikunde af. Van november 1991 tot en met mei 1997 had hij een dienstverband als assistent-in-opleiding (AIO) en toegevoegd onderzoeker bij het Interfacultair Reactor Instituut (Technische Universiteit Delft). Vanaf maart 1994 werd door hem het hier beschreven promotieonderzoek naar de analytische aspecten van speciatie van technetium in spinazieplanten uitgevoerd. Naast de jaarlijkse posterpresentaties tijdens de bijeenkomsten van de SON-werkgemeenschap Analytische Chemie in Lunteren presenteerde hij een poster op het congres "4th International Conference on Nuclear and Radiochemistry" in Saint Malo, Frankrijk (september 1996). Mondelinge presentaties werden door hem gehouden op een technetium-symposium georganiseerd door de sectie Radio- en Stralenchemie van de KNCV in Delft (juni 1995) en op het congres "Separation of Ionic Solutes 7" in Pistany, Slowakije (mei 1997).

List of Publications

Chapter 3 Harms, A. V.; Wolterbeek, H. Th.; Woroniecka, U. D. "A caveat on the use of 2-mercaptoethanol in plant technetium speciation procedures", *Applied Radiation and Isotopes* **1996**, *47*, 605-610.

Chapter 4 Harms, A. V.; Van Elteren J. T.; Claessens, H. A. "Technetium speciation: non-size effects in size-exclusion chromatography", *Journal of Chromatography A* **1996**, *755*, 219-225.

Chapter 5 Harms, A. V.; Van Elteren, J. T.; Wolterbeek, H. Th.; De Goeij, J. J. M. "A Dual Radiotracer Technique for Technetium Speciation with Emphasis on Probing of Artifacts: a Case Study for Spinach Plants", **1998**, in preparation.

Chapter 6 Harms, A. V.; Van Elteren, J. T. "Search for Plant-Induced Technetium Species with Size-Exclusion and Reversed-Phase Ion-Pair Chromatography", *Journal of Radioanalytical and Nuclear Chemistry* **1998**, *227*, in press.

Chapter 7 Harms, A. V.; Van Elteren, J. T. "Characterisation of Technetium Species in Spinach Plants using a Combination of Selective Extraction and Chromatographic Techniques", **1998**, in preparation.

Hattink, J.; Harms, A. V.; Weltje, L.; Verburg, T. G.; Wolterbeek, H. Th. "Open and Closed First-order Compartmental Models in Uptake and Speciation Studies of Micropollutants in Plants: a Case Study for Technetium-99 and Duckweed (*Lemna Minor L.*)", **1998**, in preparation.